night sky

An EXPLORE YOUR WORLD™ Handbook

DISCOVERY COMMUNICATIONS
Founder, Chairman, and Chief Executive Officer:
John S. Hendricks
President and Chief Operating Officer:
Judith A. McHale
President, Discovery Enterprises Worldwide:
Michela English

DISCOVERY PUBLISHING
Vice President, Publishing: Ann-Marie McGowan
Publishing Director: Natalie Chapman
Editorial Director: Rita Thievon Mullin
Senior Editor: Mary Kalamaras

DISCOVERY CHANNEL RETAIL
Product Development: Tracy Fortini
Naturalist: Steve Manning

DISCOVERY COMMUNICATIONS, INC., produces high-quality television programming, interactive media, books, films, and consumer products.
DISCOVERY NETWORKS, a division of Discovery Communications, Inc., operates and manages Discovery Channel, TLC, Animal Planet, and Travel Channel.

 Published in the United Kingdom by Discovery Books, an imprint of Cassell & Co, Wellington House, 125 Strand, London WC2R 0BB.

Night Sky, An Explore Your World ™ Handbook, was created and produced for DISCOVERY PUBLISHING by ST. REMY MEDIA INC.

A CIP catalogue record for this book is available from the British Library

ISBN 1 84201 003 4

Discovery Channel Online website address:
http://www.discovery.com
Printed in the United States of America on acid-free paper
First Edition 10 9 8 7 6 5 4 3 2 1

CONSULTANTS

Richard Barchfield works at Kitt Peak National Observatory near Tucson, Arizona, presenting the public observing program. A graduate of the University of Arizona, he became involved in observational astronomy while a locomotive engineer on the railroad. An accomplished lecturer, Barchfield is active in astronomy education outreach.

Louie Bernstein has worked for the last ten years as a lecturer and a specialist in the field of astronomy at the Montreal Planetarium. Bernstein is a past president of the Montreal center of the Royal Astronomical Society of Canada. Since 1995 he has written and presented *The Sky This Week*, a series of astronomical television segments for the Weather Network. The series has been nominated for several awards by the International Scientific Film Festival.

Robert Burnham was an editor for *Astronomy* magazine for many years. He is the author of several astronomy books, including *Advanced Skywatching* (Nature Compnay, 1997), *Reader's Digest Explores Astronomy* (Reader's Digest Books, 1998), and *Great Comets* (Cambridge University Press, 1999).

Wil Tirion has had a lifelong fascination with astronomical maps. In 1977 he started making his first star atlas just for the fun of it. Since then he has contributed to numerous books and magazines, including *Sky Atlas 2000.0* and *Uranometria 2000.0*, which plots more than 300,000 stars down to magnitude 9.5. Tirion lives in the Netherlands.

night sky

An EXPLORE
YOUR WORLD™
Handbook

DISCOVERY BOOKS
LONDON

Contents

SECTION ONE
6 A Celestial Tour

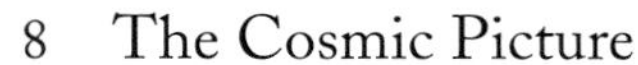

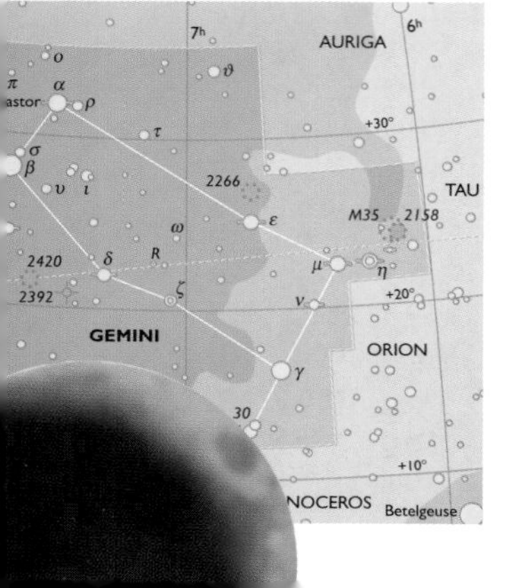

SECTION ONE

A Celestial Tour

THE COSMIC PICTURE

Discovery by discovery, astronomers are slowly unlocking the secrets of the universe and our place in it.

Despite humankind's natural inclination to assign the Earth a central role in the universe, astronomers have a different tale to tell. The humbling fact is that we reside on a tiny blue speck of a planet, just one of nine worlds that orbit an average star called the Sun. The Sun and its family of planets exist—like all other solar systems that have ever been spotted—within a vast, luminous stellar aggregation called a galaxy. Our galaxy is the Milky Way. It contains about 200 billion stars and is more than one hundred thousand light-years in diameter.

"Now my suspicion is that the universe is not only queerer than we suppose, but queerer than we can suppose."

— JOHN HALDANE
1892-1964

How much space does our little solar system occupy in the Milky Way? If the Sun and nine planets were reduced to the size of a quarter, the Milky Way would stretch four times the width of the continental United States.

Shedding Light on Light

Visible light forms a small part of the electromagnetic spectrum *(page 50)*, a type of energy that includes radio waves, infrared radiation, X-rays, and gamma rays. Each part of that spectrum is distinguished by its wavelength—radio waves are longest, gamma rays the shortest. The color of visible light is also a matter of wavelength: Red is longer than blue.

Electromagnetic waves, including light, travel through space at 186,000 miles (300,000 km) per second, fast enough to circle the Earth seven times a second. However, light is slow compared to the cosmic distances it travels. Astronomers measure these distances in terms of light-years. One light-year is equal to 5.87 trillion miles (9.45 trillion km)—the distance that light travels in a year. It takes sunlight about eight minutes to reach the Earth, and about 4.3 years for light to arrive from Proxima Centauri, the next closest star.

Since light takes time to travel across space, we see distant objects as they were in the past. When we look, for example, at the star Betelgeuse in the constellation Orion, which is 427 light-years from Earth, what we are seeing is the star as it was 427 years ago—not as it is today.

Many celestial objects are not just hundreds, but millions, of light-years away. For reasons of practicality, astronomers use distance units called parsecs in their calculations. One parsec is equal to 3.26 light-years.

But this is only the beginning. The Milky Way is one of about thirty galaxies that form a small cluster known as the Local Group. The Local Group, in turn, is one of dozens of small clusters centered on a large collection of more than twenty-five hundred galaxies called the Virgo cluster.

And still there is more. Altogether, these galaxies and galaxy groups form what is called the Local Supercluster. And the universe contains millions of such superclusters.

FROM LARGE SCALE TO SMALL—AND BACK

At the opposite end of the scale, cosmic structure consists of atoms. Indeed, everything in the universe is made of atoms: people, planets, stars, and galaxies. These building blocks are unimaginably small: about ten thousand billion could fit in the period at the end of this sentence. Atoms have their own structures. Each one consists of a nucleus of fundamental particles called protons and neutrons, and this nucleus is surrounded by orbiting clouds of negatively charged electrons. Despite its seemingly solid appearance, our material world is mostly empty space: If the nucleus of an atom was the size of a grape, the electron clouds would be a half mile away. The same sort of emptiness extends to our solar system. If our Sun were a grape, Earth would be but a speck of dust orbiting nine feet (3 m) away. Pluto, the most distant planet, would be the distance of a football field away.

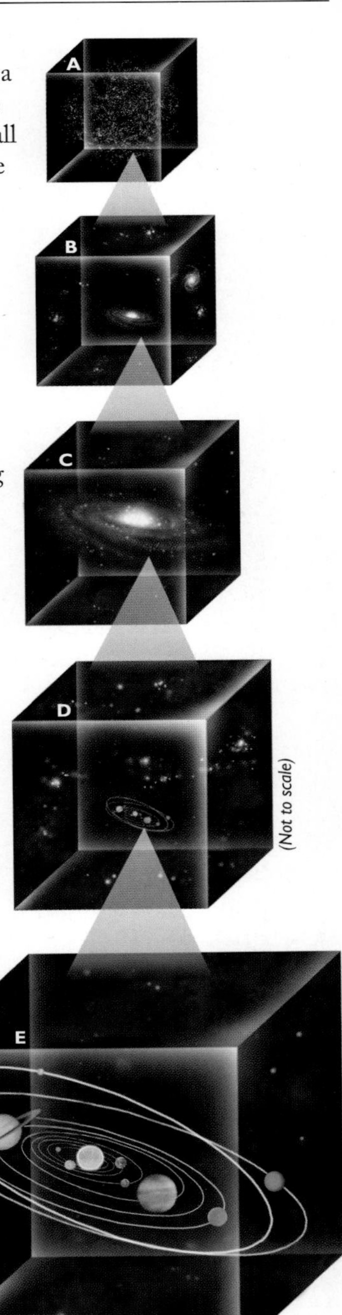

You are here

Zooming in on hundreds of superclusters—a collection of thousands of galaxies known as the Virgo cluster (A)—reveals the Local Group (B), an assembly of thirty or so galaxies, including our Milky Way (C). In one of the outer spiral arms of the Milky Way resides the Sun and its nearby celestial stellar neighbors (D). Revolving around our star are nine planets (E). We live on the very small blue one, the third planet from the Sun.

The Expanding Universe

The most distant object visible to the naked eye—2.3 million light-years away—the Andromeda Galaxy played a central role in helping astronomers determine that the universe is expanding.

When the great theoretical physicist Albert Einstein devised his general theory of relativity in 1915 *(page 14)*, he realized that it made a rather disturbing prediction: The universe was expanding. At that time, most astronomers believed that the universe had always existed. But an expanding universe meant a cosmic beginning, and that in turn conjured up questions of what had existed *before* that beginning.

The previous year, American astronomer V. M. Slipher had measured the velocities of what were though to be twelve nearby "spiral nebulae," clouds of interstellar dust and gas, and found that they were receding from the Earth.

Slipher made his calculations on the basis of the Doppler effect: the change in the wavelength of light or sound as the emitting source moves toward us or away. Think of a train passing by. As it approaches, the pitch of its whistle rises as the sound waves are compressed. Once it passes

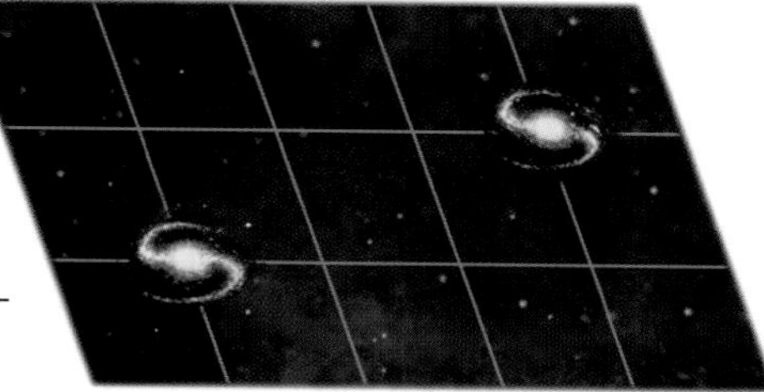

Expanding space

The simplest way to visualize cosmic expansion is to imagine galaxies placed on a tiled floor. As the tiles expand and the floor becomes larger, the galaxies move farther apart.

by, the waves are stretched and the pitch drops. The same holds true for light: When a light source approaches us, its light becomes bluer because its wavelength gets shorter; as the source recedes, its light is reddened because the wavelength gets longer—an effect known as redshift.

In 1924 another U.S. astronomer, Edwin Hubble, began observing spiral nebulae with the recently completed hundred-inch Mount Wilson telescope. Hubble was able to detect individual stars in the "Andromeda nebula" and prove that, in reality, they were part of an island of stars more than one million light-years away. The Andromeda nebula had become the Andromeda Galaxy.

Subsequent observations of other spiral nebulae and their redshifts led Hubble to discover that the Milky Way was surrounded by galaxies, and that these galaxies were all moving away. The farther the galaxies were, the faster they were receding. This became known as Hubble's Law.

In fact, there was no reason to believe that the Milky Way galaxy was the only stationary one. The view from every other galaxy would be the same—other galaxies receding. Much to Einstein's shock, the universe *was* expanding.

According to Einstein's equations, however, the galaxies themselves are not moving through space; the space between galaxies is expanding. Furthermore, if space is expanding, then at some time in the past the universe was compressed into a single, very dense state. Measuring the rate of cosmic expansion allows astronomers to calculate the age of the universe, currently estimated at twelve to fifteen billion years.

From Hubble's Law came the Hubble constant, a measure of the rate of galactic recession, hence of cosmic expansion. Ironically, the Hubble constant has been anything but constant. It is continually being revised as new and more accurate observations are made. The current estimate is forty-three miles (70 km) per second per megaparsec. (One megaparsec is about 3.26 million light-years.) In other words, a galaxy one megaparsec away recedes at a speed of forty-three miles (70 km) per second. A galaxy twice as far recedes twice as fast, and so on.

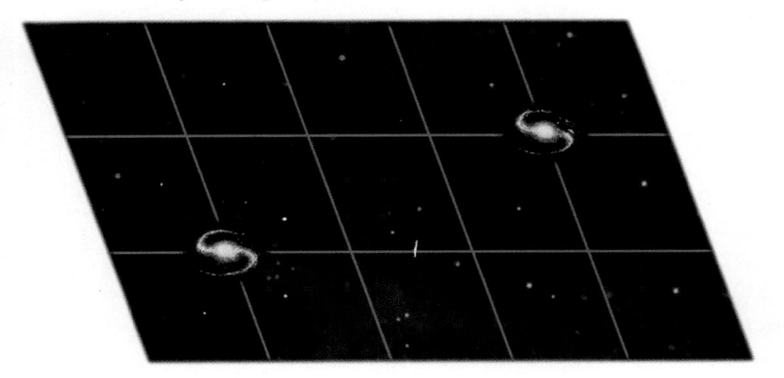

The Case for the Big Bang

Fathoming the origins of the universe may sound like an exercise in pure speculation. But clues to the mystery of creation lie all around us. Though they are subtle, they can be brought into focus with the right combination of observation and creative thinking.

Today most cosmologists believe the universe began in a sudden expansion of space, time, matter, and energy that they call the Big Bang. Their main supporting evidence comes from the expansion of the universe and the recession of the galaxies *(page 10)*: Everywhere we look, galaxies are moving away from us.

Only two possibilities could account for such an expansion: Either everything started moving apart from a common point of origin—the Big Bang—or everything is continually being pushed aside as new material comes into existence between the galaxies, a notion known as the Steady State theory.

In the Beginning
For the first three hundred thousand years after the birth of space and time, the universe was opaque. With the creation of atoms from protons and electrons, radiation could travel freely and the universe became visible.

If the Big Bang really did occur, then we should see evidence of it today. A good theory produces good predictions, and the Big Bang theory predicts that as the fires of creation expanded and cooled, a cosmic "afterglow" would have filled the universe. This afterglow would take the form of microwave radiation at a temperature of -453 degrees Fahrenheit (-273°C)—just three degrees above absolute zero, the temperature at which all molecular motion would stop. Evidence of this radiation turned up in 1965, when Arno Penzias and Robert Wilson, two researchers at Bell Telephone Labs in Holmdel, New Jersey, announced that they had discov-

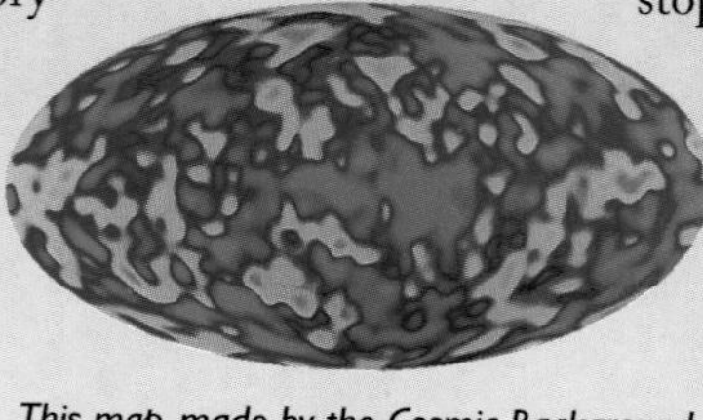

This map, made by the Cosmic Background Explore (COBE), shows tiny temperature variations in the "fading light" of the Big Bang. The pink regions are thirty millionths of a degree "hotter" than the blue ones.

ered just such a remnant of the birth of the universe. Score one point for the Big Bang.

Another prediction of the theory is that as protons, neutrons, and electrons began to condense out of the primordial fires of creation, three atoms of hydrogen would have formed for every atom of helium.

Recent measurings of the relative abundance of these elements show that there are, indeed, about three hydrogen atoms for every helium atom in the universe. Score another point for the Big Bang.

Still, the Big Bang theory has met some difficulties. The theory predicts a very smooth background of energy—too smooth to account for the rapid clumping of matter that we see today in the form of galaxies. But that problem was resolved in 1995, when observations taken by the Cosmic Background Explorer satellite *(opposite)* revealed variations in the early distribution of energy, enough to account for the rapid formation of galaxies (matter and energy being equivalent).

Another difficulty with the Big Bang was that some of scientists' best measurements revealed stars that were older than the universe itself—something clearly impossible. Recently, though,careful observations of distant exploding stars have indicated that the expansion of the universe is actually speeding up. If the expansion is accelerating, then it must have been expanding more slowly in the past, immediately after the Big Bang. That means the universe is somewhat older than scientists had previously believed—in fact, about twelve to fifteen billion years old, old enough to contain the oldest stars.

But all is not settled yet. Both the Big Bang and the Steady State theories require the creation of something from nothing. And that is one mystery that still baffles scientists.

A Matter of Gravity

The entire universe is dominated by just four forces, which orchestrate the interplay of space, time, energy, and matter. The strong and the weak forces control nuclear interactions within atoms. The electromagnetic force and gravity determine how individual atoms and large masses interact.

Mass gives rise to gravity; the more mass, the more gravity. This profoundly influences the future of the universe. Not enough mass and the universe will expand forever at an increasing rate. Such a universe is said to be open. More than enough mass and gravity will eventually stop the expansion, causing everything to fall back in a "Big Crunch." This universe is said to be closed. Just the right amount of mass and the universe goes on expanding forever, but increasingly slowly. In this scenario, the universe is flat, perfectly balanced between open and closed.

The Ups and Downs of Gravity

Gravity is everywhere. It keeps our feet planted firmly on the ground, it holds the planets in orbit around the sun, and it prevents the stars and galaxies from flying apart. In short, gravity helps to hold the universe together. And yet it is the most enigmatic of nature's forces.

The renowned English scientist and mathematician Sir Isaac Newton understood gravity as the mutual force of attraction between objects. Though Newton correctly determined that gravity increases with mass and decreases with distance, he never figured out what causes the force in the first place.

At the turn of the twentieth century, the Russian-born mathematician Hermann Minkowski suggested that the universe actually consists of four dimensions: the familiar three dimensions of space plus one dimension of time. This concept, known as space-time, captivated a young contemporary named Albert Einstein and helped him form the basis of a new theory of gravity. According to Einstein, the presence of mass produces a curvature in space-time, and it is this curvature that we perceive as gravity.

Try to imagine space-time as a rubber sheet stretched taut. Any object placed on this sheet will produce a depression. When a smaller object is placed nearby, it rolls into the depression, "gravitating" toward the larger object. The more massive the object, the larger the depression and the greater the gravitational attraction. In this analogy, an orbit is achieved when the smaller object circles the larger one with enough velocity to counterbalance the pull of "gravity:" Too slow and it spirals in; too fast and it spirals out.

A black hole (*page 28*) can then be visualized as an extremely massive object embedded in the rubber sheet. The resulting "gravitational well" is so deep that even light is too slow to escape its pull.

Gravity's effect on space and time
Gravity is not merely a force that operates between two bodies, pulling them together; it also affects both space and time. All objects cause curves in so-called spacetime—shown at left as a taut rubber sheet; the greater the mass, the greater the distortion. In the case of a black hole (far left), *time is affected and eventually stops as the body enters the hole. Depressions in space-time also bend light* (near left), *causing an observer (A) looking at an object (B) in the distance to perceive that it is actually in a different location (C). In fact, the light is bent by a massive body in the foreground.*

Einstein had another startling suggestion to make, based on the fact that the speed of light (C) is constant. This suggestion took the form of the equation $E=MC^2$, which says that energy and mass are different states of the same thing. Einstein concluded that if mass is affected by gravity, then energy should as well. For instance, light, a form of energy, should be bent by gravity.

Many doubted the conclusion, but the proof for Einstein's seemingly wild claim came in 1919 during a total eclipse of the sun. At the moment when the sun's sphere was blocked out by the moon, scientists observed that several stars near the edge, or limb, of the sun had their apparent positions altered by the sun's gravity. The amount of that deflection precisely matched Einstein's predictions.

That bending of light has provided some practical benefits for astronomers in their efforts to understand the universe. When light from a distant object is bent around a massive foreground body, such as a galaxy, the image is stretched and distorted. The resulting effect, called gravitational lensing, causes a magnification of the distant object revealing information about both bodies.

A QUESTION OF TIME

Another even more surprising consequence of Einstein's theory is that time itself is affected by gravity. Clocks that are closer to a gravitational field run slower than those farther away. The frequency of light is a kind of clock: It is the number of wavelengths that pass a given point per second. When light passes through a strong gravitational field, its frequency slows down and its wavelengths get longer. This results in the light being shifted toward the red, longer-wavelength end of the spectrum. Astronomers use this effect, known as gravitational redshift, to measure the strength of gravitational fields and the mass of the objects producing them.

Einstein also predicted another phenomenon known as gravity waves—"undulations" in space-time that travel at the speed of light. Gravity waves are caused by sudden catastrophic events, such as exploding stars or colliding black holes, and are similar to the ripples that form when a pebble is thrown into a pond. Gravity waves are extremely weak and have not yet been directly observed, but projects such as the LIGO observatory *(page 60)* promise to change all that, revealing cosmic processes in a whole new "light."

The Emergence of Galaxies

The vast, starry islands we call galaxies are among the largest structures in the universe—and the most beautiful. But how did they form? The answer lies back at the beginning of time.

The first galaxies began to form one billion years after the Big Bang, very early in "cosmic time." Until recently, astronomers were hard put to explain how such enormous structures could have emerged so quickly. The reason, it turns out, begins with the Big Bang.

"Now entertain conjecture of a time / When creeping murmur and the poring dark / Fills the wide vessel of the universe"

— WILLIAM SHAKESPEARE
Henry V

In the first brief instant after creation, the universe was aburst with primordial energy. As the universe expanded and cooled, gravity began to manifest itself. Then the infant universe entered a short period of violent expansion—inflation, as physicists call it. Subtle fluctuations in the primordial energy were greatly magnified as the universe inflated. These fluctuations later became "furrows" along which galaxy clusters would form.

Inflation lasted but an instant, then the universe resumed the normal rate of expansion we see today. One second after the beginning, as the universe continued to expand and cool, protons, neutrons, and electrons condensed out of the background energy in greater numbers where energy was slightly "denser." One hundred seconds later, protons and neutrons combined to form helium nuclei. Hydrogen nuclei, which are simply protons, already existed. So, in less than two minutes, all the subatomic elements of nature had been created.

About three hundred thousand years elapsed before the universe had cooled enough for electrons to be captured by hydrogen and helium nuclei to form atoms. These atoms slowly gravitated into vast filamentary clouds, from which the galaxies would soon emerge.

The Stuff We're Made Of
The building blocks of nature are as old as the universe itself. Virtually all the protons, neutrons, and electrons that exist today—including those that make up our bodies—were formed by the time the universe was a mere hundred seconds old.

GALAXIES BEGIN TO APPEAR

One billion years after the Big Bang, clouds of hydrogen and helium began clumping under the force of gravity. As the clouds grew, nascent galaxies, or protogalaxies, began to form. The universe was smaller then, and the protogalaxies were closer, which promoted an intense flurry of interaction. Many smaller clouds condensed gravitationally out of larger, more tenuous ones; others merged with neighboring clouds.

At the same time, protogalaxies were growing as a result of infalling hydrogen and helium. The more massive the protogalaxies became, the more gas they attracted. The interaction between the clouds, combined with their individual motions, eventually initiated a slow rotation.

As the clouds collapsed further under the force of gravity, some spun faster becoming disklike; others retained a more or less elliptical shape. Eventually these primal galaxies acquired enough mass for stars to ignite within them, and the universe took on its appearance today.

Like islands on Earth, galaxies are clustered in archipelagoes, which are embedded in huge clouds of gas. Such clusters and intergalactic gas extend in filamentary structures that are hundreds of millions of light-years long and englobe vast galactic voids hundreds of millions of light-years across.

This remarkable image, called the Hubble Deep Field, reveals about fifteen hundred galaxies, seen in an area of sky no larger than a grain of sand held at arm's length.

A Gallery of Galaxies

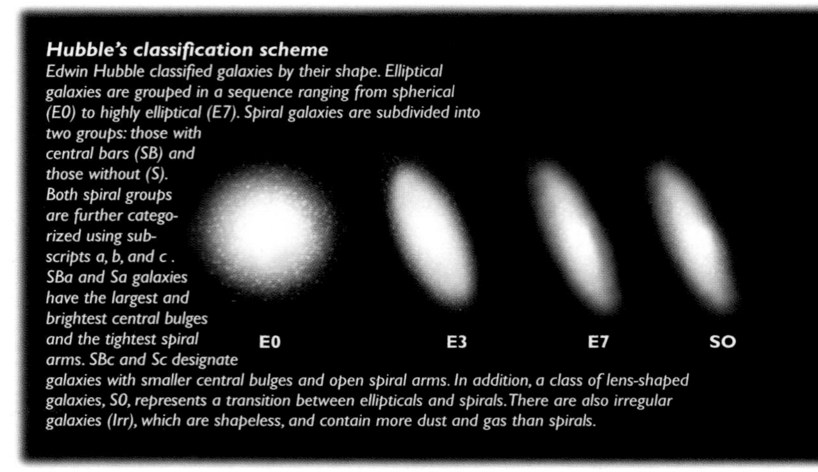

Hubble's classification scheme
Edwin Hubble classified galaxies by their shape. Elliptical galaxies are grouped in a sequence ranging from spherical (E0) to highly elliptical (E7). Spiral galaxies are subdivided into two groups: those with central bars (SB) and those without (S). Both spiral groups are further categorized using subscripts a, b, and c . SBa and Sa galaxies have the largest and brightest central bulges and the tightest spiral arms. SBc and Sc designate galaxies with smaller central bulges and open spiral arms. In addition, a class of lens-shaped galaxies, S0, represents a transition between ellipticals and spirals. There are also irregular galaxies (Irr), which are shapeless, and contain more dust and gas than spirals.

Like snowflakes, no two galaxies are exactly alike: Each has its own unique appearance. But because galaxies formed under a limited set of conditions, they have certain features in common, which allows them to be grouped in general categories. Although many classification schemes exist, a system developed in 1925 by the astronomer Edwin Hubble is most widely used *(above)*.

Hubble categorized galaxies according to their shape: elliptical, spiral, barred spirals, and irregular. About 78 percent of all large galaxies are spirals, and half of those have bars; 18 percent are elliptical, and 4 percent are irregular. Though it is tempting to think that spiral galaxies gradually wind themselves up, becoming tighter and changing from one form to another, this is not the case. In fact, the shape of all galaxies is determined at birth and remains relatively stable, unless affected by gravitational interactions and collisions with their neighbors.

Compared to spiral galaxies, ellipticals rotate very slowly and exhibit a smooth, symmetrical structure with no spiral arms and little dust or gas. In fact, virtually all the dust and gas in elliptical galaxies was consumed billions of years ago in a rapid flurry of star formation. As a result, virtually no new stars are being created in these galaxies: They consist entirely of old stars containing hydrogen and helium and little else.

Spiral galaxies, in contrast, rotate relatively rapidly, and their disks contain large amounts of dust and gas that coalesce to create star-forming regions. These regions seed the spiral arms with hundreds of blue stars that give the disks a bluish hue. The bars and central bulges are densely populated with older stars containing hydrogen, helium, and many heavier elements.

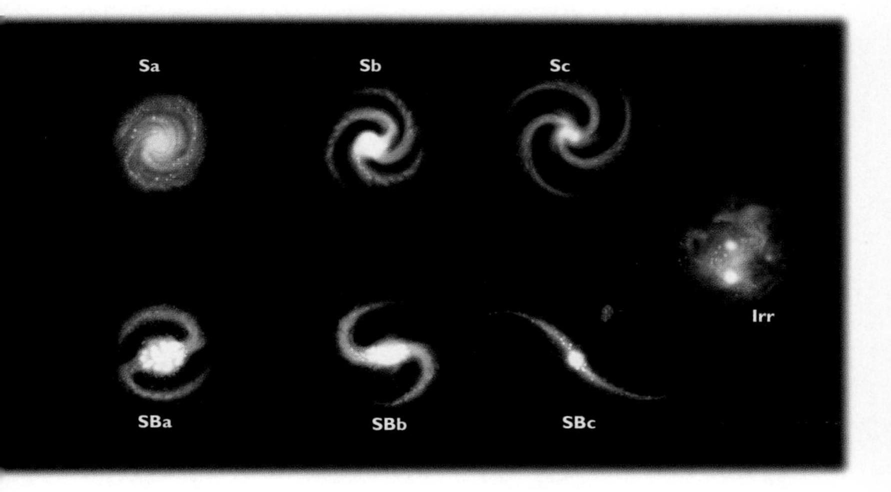

ACTIVE GALAXIES

There are close to one hundred billion galaxies in the universe. Roughly 1 percent of them are classified as active galaxies, powered by supermassive black holes at their centers. These include quasars (quasi-stellar objects), the most luminous objects in the universe.

Quasars generate hundreds of times more radiation than an entire ordinary galaxy—all from a tiny nucleus just a few light-days across. They were more common in the distant past, at a time when galaxies were packed closer together and thus were more likely to collide. Collisions resulted in large-scale galactic mergers that fed millions of stars into their central black holes, which generated hundreds of times more energy than the galaxies would have produced from the normal thermonuclear reactions that power the stars.

Invisible Matter

Spiral galaxies rotate so quickly that their stars should fly off into space. And entire galaxy clusters behave as if they contained more mass than we actually see.

How can we explain these mysteries? Scientists think that the answer lies with "dark matter." Only a small part of the universe, they say, emits electromagnetic radiation; the rest lies hidden from view.

Astronomers are now trying to account for this dark matter. Their explanations range from fast-moving elementary particles that were left over from an earlier cosmic epoch to the subatomic, almost massless particles that are known as neutrinos *(pages 60 to 61)*.

Yet this dark matter is crucial to the future of the universe. More dark matter means more gravity, which will slow the rate of expansion of the universe. Without enough dark matter, the universe will continue to expand forever *(page 12)*.

The Milky Way

On a clear, dark night in the country, far from the light-polluted city skies, thousands of stars are visible, glittering against the backdrop of space. Yet this cosmic panorama, as vast as it may seem, represents a very small part of our galaxy. If the sky is dark enough, you can also see a diffuse band of light stretching overhead. That band is the starry disk of our galaxy, seen edge-on from within. It contains billions of stars that are too far away to discern with the naked eye. Instead, their light blends with interstellar dust and gas to form a glowing haze.

Over the ages, this glow has been given many names by different civilizations. The ancient Egyptians referred to it as the Road of Souls; the Kung bushmen of the Kalahari deemed it the Backbone of Night. Modern astronomers call it the Milky Way galaxy, our island-home in the universe.

The Milky Way is a vast disk-shaped system of stars embedded in a large halo peppered with stars and ancient, spherically shaped globular clusters.

The disk itself is thin and tapered, with a bulge at its center, and contains large clouds of interstellar dust and gas that coalesce into pink-colored, star-forming regions. These replenish the spiral arms of the galaxy with fresh families of hot, young, blue stars, known as open or galactic clusters—groupings of stars that are more

The Milky Way contains more than 200 billion stars, including our sun.

spread out than the more tightly concentrated globular clusters. The Milky Way has hundreds of star-forming regions, and about twelve hundred known open clusters.

RECENT DISCOVERIES

All this dust and gas obscures our view along the galactic plane, which makes it difficult to determine the galaxy's exact shape. However, modern instruments have penetrated the haze to some extent, revealing some surprising features.

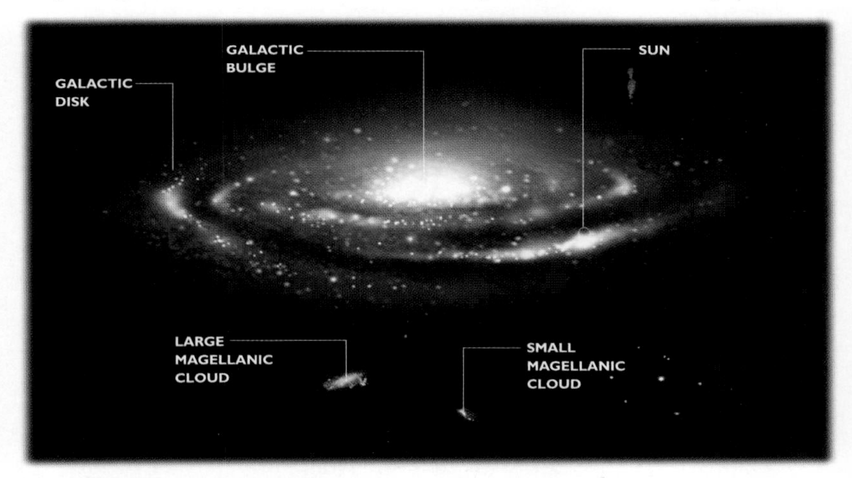

The Milky Way and its Neighbors

The Milky Way stretches 120,000 light-years across with an average thickness of 4,000 light-years. Toward the bulge, the disk thickens to 12,000 light-years. The Sun shown here greatly magnified, is located in one of the outer arms of the galaxy. The Milky Way's closest neighbors—actually satellites of our galaxy—are known as the Large and Small Magellanic Clouds, 165,000 and 195,000 light-years distant respectively. The Magellanic Clouds are visible as hazy patches in the night sky of the southern hemisphere.

For one, the Milky Way was long thought to be a typical, Hubble-type Sb spiral galaxy *(pages 18 to 19)*, similar to M31 (the Andromeda galaxy). However, recent observations show that the central bulge is somewhat bar-shaped. This means that our galaxy is most likely an intermediate Sb/SBb barred spiral.

Moreover, recent studies indicate that the Milky Way is much more active than we once imagined. Intense X-rays and plumes of gamma radiation blaze from its core, where stars whip around an unseen central region at over 560 miles (900 km) per second—strong evidence of a supermassive black hole hidden in the galactic core.

In addition, the Milky Way is in the process of cannibalizing its smaller neighbors. Two nearby irregular galaxies, the Large and Small Magellanic Clouds, are being gravitationally consumed as the Milky Way tears streamers of dust and gas away from them. Astronomers estimate that within another ten billion years, the Milky Way will have swallowed all the matter in the clouds and the two neighboring galaxies will no longer exist.

The Evolution of Stars

Emerging in cosmic clouds of gas and dust and later fueled by nuclear fusion, stars live out a life cycle that may last tens of billions of years. The end, when it comes, is determined from the moment of birth.

From earliest times, humans have gazed up at the night sky and wondered about the stars. Glittering like gems in their celestial settings, those pinpoints of light once seemed immutable and serene, and far removed from our world. In truth, they are violent and subject to extreme change.

Like people, stars are born, they live out their lives, and they eventually expire; and like people, stars come in different sizes and colors. But the links of stars to life are not merely metaphorical. At the end of their span of existence, some stars explode, creating the very atoms that all living matter is made of.

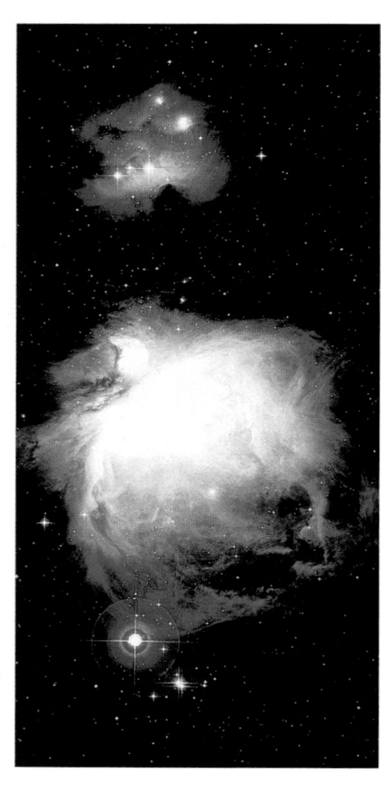

"Look at the stars!" Look, look up at the skies / O look at the fire-folk sitting in the air!"

— Gerard Manley Hopkins
The Starlight Night

Stars are born within the vast, cool clouds of molecular dust and gas that abound by the thousands, within the disks of spiral galaxies like our own Milky Way. These molecular clouds consist of hydrogen, helium, and traces of other elements. Over a period ranging from ten thousand years to ten million years, depending on the size of the star, areas within these clouds slowly begin to collapse and rotate, as the force of gravity pulls the dust and gas together.

The initial collapse is triggered when disturbances

such as nearby exploding stars or passing galaxies produce shock waves that compress the clouds. As the density of these areas gradually increases, they collapse into what are called protostars. As more material spirals inward, the newly formed protostars are surrounded by revolving, disk-shaped clouds. Under increasing pressure and temperature, the nascent star begins to glow, emitting light, heat, and radio waves.

BURNING BRIGHT

As the protostar's core continues to collapse, its temperature rises, aided by the surrounding cloud, which traps heat and acts as an insulator. Gradually, the protostar's core temperature reaches about seven million degrees Fahrenheit (3,900,000°C)—hot enough for nuclear fusion to take place. The time it takes for this to happen depends on the mass of the star, ranging from as much as one billion years for a low-mass red dwarf to ten thousand years for a massive blue giant. In the case of our sun, this period lasted for about ten million years.

Up to this point, the protostar's heat was generated by the force of gravity squeezing the core. But now, nuclear fires ignite the stellar core as hydrogen atoms combine to form helium, releasing a prodigious flow of energy in the process. And so, a new star is born.

But the process of star birth comes with a price. When infant stars "turn on," the tremendous outward flow of their radiation prevents them from accumulating more mass and growing; and as the newborn stars "light up," their combined radiation gradually blows away the cloud in which they formed. This in turn prevents more new stars from forming. Before the clouds actually clear, the blazing stellar nursery is revealed in all its glory, like the Great Orion Nebula *(opposite)*.

Visible to the naked eye in the constellation Taurus, the Pleiades (M45) (above) *are a collection of three hundred to five hundred stars roughly fifty million years old—very young in astronomical terms. The light from these recently formed hot, young, blue stars reflects off the remaining wisps of the interstellar dust in which they formed. The central region of the Great Orion Nebula (M42)* (opposite) *glows from the intense effects of radiation caused by the recent birth of stars at its center.*

The Life Cycle of Stars

Stars brought to life by the collapse of swirling clouds of dust and gas *(pages 22 to 23)* may vary in size from about one-tenth of the sun's mass to about a hundred times that. Stars much more massive than a hundred suns are believed to be unstable, while much smaller masses than a tenth of the sun will not give rise to the high temperatures needed for nuclear fusion. Instead, these so-called brown dwarfs generate their heat by gravitational compression, until they sputter out, the dying embers of a failed star birth.

The ignited stars, meanwhile, will spend most of their lives quietly converting the hydrogen fuel in their cores to helium. The star's mass determines its longevity. Although more massive stars have more fuel to burn, the increased pressure and temperature cause their burn rates to be proportionately higher, so—ironically—their lifetimes are shorter than those of less massive stars. The most massive stars will burn hydrogen for only a few million years, while the least massive may burn for more than a trillion years. Our Sun was born about five billion years ago and will live for another five billion years.

LOSING HYDROGEN

No matter what the mass of the star, the hydrogen fuel burning in the core is eventually exhausted. As the hydrogen fusion falters, the radiation pressure stabilizing the

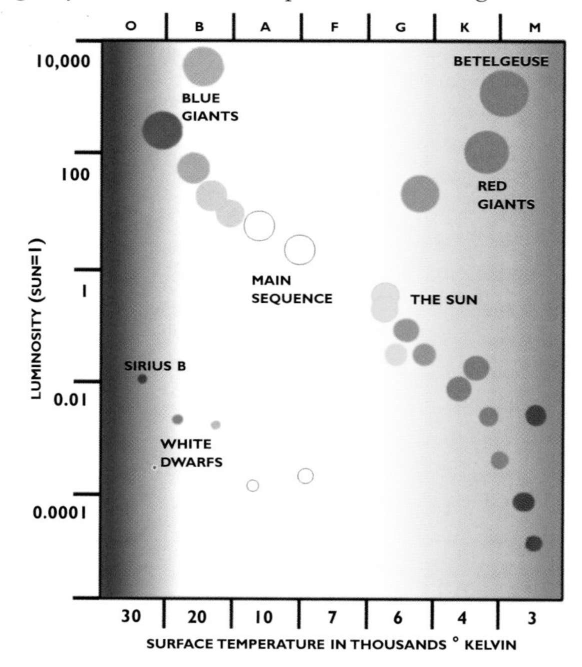

A Revealing Chart
There is a link between a star's temperature and color, and its luminosity, as shown in the Hertzsprung-Russell (H-R) diagram at right. Hot, bright blue giant stars are in the upper left, while cooler, dim, red dwarf stars are in the lower right. Most stars, including our Sun, fall on a diagonal line, known as the main sequence, which describes the hydrogen-burning phase of a star's life. When stars run out of hydrogen in their cores, they leave the main sequence, becoming red giants as they begin burning helium. Eventually, red giants collapse and heat up becoming white dwarfs before they die out. Stars are classified according to their color and temperature with the letters OBAFGKM: O being blue hot and M, cool red. Star temperatures are measured in Kelvins. To convert Kelvins to Fahrenheit, add 459° (for Celsius, add 273°).

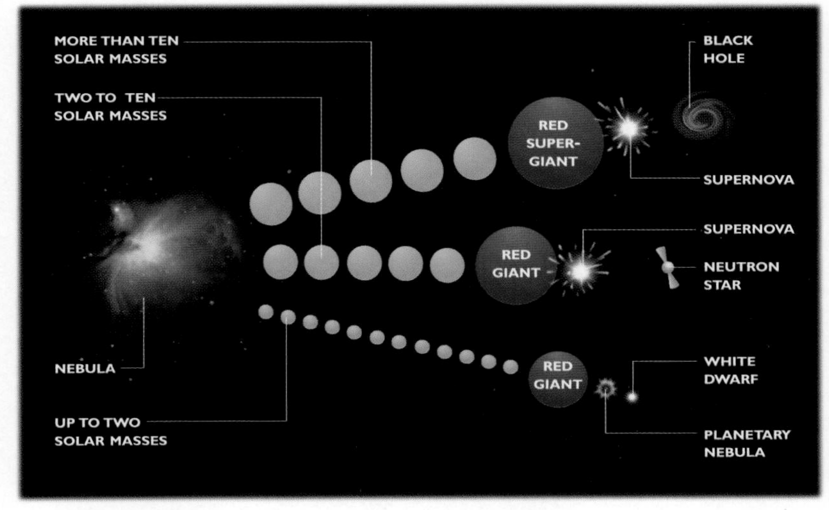

From Birth to Death

The way a star dies depends upon its mass at birth. From its birth in a collapsing cloud of nebular material, a star with a mass about the same as our Sun (bottom) *will exhaust its fuel in ten billion years and swell into a red giant before becoming a white dwarf. A star of about ten solar masses will become a supernova, exploding and then collapsing into a neutron star. Stars of greater mass turn into red supergiants and then collapse into an area so dense that light cannot escape—a black hole.*

core subsides and gravitational contraction resumes. The heat produced, together with the availability of fresh unfused hydrogen just outside the core, leads to a hydrogen-burning "shell" surrounding the collapsing core. The star's outer layers swell enormously and cool, and the gaseous ball becomes a red giant, a behemoth more than a hundred times its previous size.

The larger size more than makes up for the cooler surface temperature, and the star's total power output and luminosity increases dramatically. When our sun reaches this stage, its outer layers will extend past Mercury's orbit, vaporizing that planet, and the sun's energy output will increase a thousand times.

Eventually, the star's core, now helium, collapses and heats enough to allow helium nuclei (two protons and two neutrons) to fuse, forming carbon and oxygen nuclei. This secondary fusion process, which occurs only when the core's temperature rises to about eighteen million degrees Fahrenheit (10,000,000°C), again provides radiation pressure that stabilizes the core against collapse. Well-known helium-burning red supergiants include Betelgeuse, which is easily visible to the naked eye as the ruddy, reddish orange "left shoulder" in the constellation Orion.

If the star is between ten and twenty times more massive than the sun, this phase may include a period of instability: The star regularly expands and contracts, and cools and heats. Such a star is called a Cepheid variable, named after the fourth-brightest star in the constellation Cepheus, which varies in brightness by nearly a factor of two every five days.

NO MORE HELIUM

The star's reprieve after it begins to burn helium in its core is only temporary, however, since the helium fuel is also soon exhausted. A star like our Sun will spend only about a billion years in the core helium-burning phase (compared to about ten billion years in the core hydrogen-burning phase). For stars less than about ten times the sun's mass, the helium-burning phase is the last stage in core nuclear fusion. As the helium is exhausted, the core again resumes its collapse, but helium burning continues in a shell outside the star's central region. Again, the star balloons in size and increases in luminosity; our sun will reach nearly to the Earth during this phase, engulfing Venus. The star ejects mass from its outer layers. Meanwhile, the core, now composed of carbon and oxygen (the products of helium fusion) shrinks. The density rises to fifty tons per teaspoonful. The collapse only stops when the electrons cannot be packed any tighter. As the outer layers of the dying star are cast off, this ultra-dense core is exposed. It is hot—forty-five thousand degrees Fahrenheit (25,000°C)—but small, no bigger than Earth, and so is called a white dwarf.

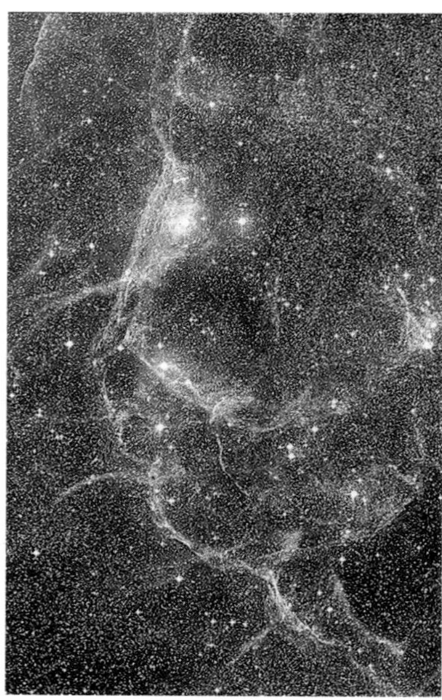

This diaphanous glow is what remains of a star that exploded twelve thousand years ago. In just a few seconds, the Vela supernova would have generated as much energy as an entire galaxy's stars shining for a year.

Stars with more that about ten solar masses, undergo a similar evolution to a carbon/oxygen core. However, the more the massive core develops, the higher the temperatures and pressures that cause additional phases of nuclear burning: carbon fusion to neon and magnesium, then on to silicon and sulfur, and finally silicon-burning to produce iron. The fuel for each stage is the "ash" of the preceding stage, and each stage requires a higher temperature to ignite.

At the conclusion of each stage, the core collapses and heats as the fuel is exhausted, and fusion continues in a shell outside the core.

The star resembles a giant onion, with successively deeper layers burning hydrogen to helium, helium to carbon, carbon to neon and magnesium, and neon and magnesium to silicon, with the core burning silicon to iron at about five billion degrees Fahrenheit (2,800,000,000°C). Each successive core-burning stage powers the star for a shorter period of time; a hundred-solar mass star will exhaust its core silicon in only about one day.

THE END OF THE ROAD

Iron, however, is the end of the road. The star can't burn iron to produce heavier nuclei—such processes require more energy than is available. The iron core collapses by a factor of one hundred in just a few seconds, with densities reaching a million tons per teaspoonful. The electrons and protons are compressed so closely that they merge to form neutrons and neutrinos—electrically neutral, nearly massless, subatomic particles.

Meanwhile, the unsupported outer layers of the star fall inward, but rebound off the dense nuclear material in a shock wave. The neutrinos pass through the shock wave, which must work its way out through the outer layers of the dying star. The shock wave literally blows the star apart—a spectacular event called a supernova. The explosion blazes briefly with the light of billions of stars, then subsides in about a year.

In 1987 light from a supernova explosion in a neighboring galaxy, the Large Magellanic Cloud, reached Earth and caused tremendous excitement among scientists. For the first time, the neutrino pulse was detected in large underground observatories, confirming the essential details of this supernova scenario. Since 1987 more detectors have been brought into operation and astrophysicists are excitedly awaiting the next supernova in our galaxy *(pages 60 to 61)*.

The supernova leaves behind the core of the now-dead star. Depending on its mass, this last vestige of the star can take one of two forms: a neutron star or a black hole *(pages 28 to 29)*, both among the most mysterious objects in the universe.

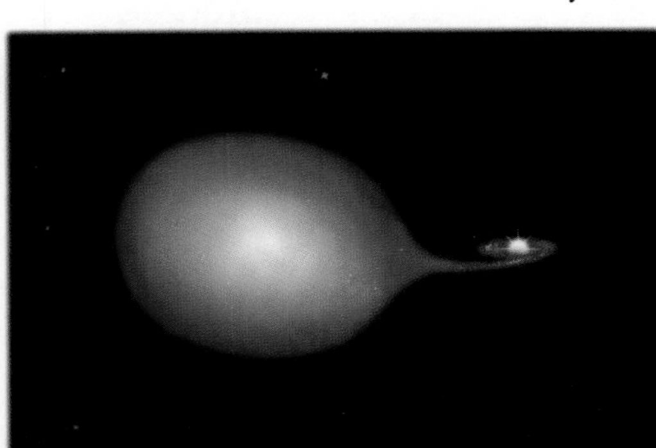

Close neighbors
More than half the stars in the universe are binaries, two stars that orbit around each other. If one of the stars becomes a white dwarf and the other remains on the main sequence (page 24), *the white dwarf can actually tear matter away from its companion's surface and flare up into a nova, increasing its brightness by ten times or more.*

Neutron Stars and Black Holes

A long time ago, people believed that stars were actually tiny pinholes in the sky, through which the fires of creation shone. Today we know that stars are really blazing globes of hydrogen fueled by thermonuclear reactions at their cores. Yet, in some poetic way, the ancient belief is valid. The fires of creation do burn at the hearts of supergiant stars, as they transform hydrogen and helium into the elements that constitute our world; and some of these stars do, indeed, end their lives by becoming "holes in space."

Neutron stars and black holes represent the end stage in the evolution of supergiant stars—stellar behemoths that can weigh as much as a hundred suns and measure more than four hundred solar diameters. When these giants run out of fuel, they collapse and explode into a supernova, leaving behind a cloud of dust and gas, and a very small, highly compressed stellar core.

MASS MATTERS

What happens next is determined by the core's mass. In 1939 Indian astrophysicist Subrahmanyan Chandrasekhar discovered that when stellar cores have about 1.4 times the sun's mass, the force of gravity compresses their atoms so much that electrons and protons combine to form neutrons. This mass is known as the Chandrasekhar limit, and the result is a neutron star. Neutron stars cram the mass of nearly 1.5 suns (or 467,000 Earths) into a core just fifteen miles across—about the size of most major cities.

Neutron stars have a surface temperature of about two million degrees Fahrenheit (1,100,000°C) and they radiate X-rays, gamma rays, and a small amount of visible light. They also have extremely powerful magnetic fields that beam radio waves along the star's magnetic poles. Neutron stars rotate very rapidly—up to a few hundred times per second—and since their magnetic poles rarely coincide with their rotational axes, the radio beams can sweep past Earth like a lighthouse beacon, producing radio pulses. This sort of stellar emitter is called a pulsar *(below)*.

BLACK HOLE MAGIC

If the post-supernova, stellar core is more massive than two or three suns, it exceeds the Chandrasekhar limit and continues to collapse, finally disappearing altogether. The stellar core becomes an infinitely small,

The Sweeping Beam of a Pulsar
Neutron stars often rotate rapidly, emitting a beam of radio energy that strikes Earth once per rotation. They are known as pulsars. Today about a thousand pulsars have been detected out of an estimated hundred million neutron stars in our galaxy.

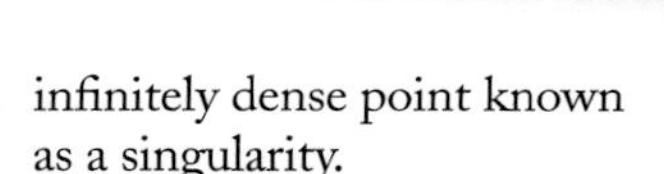

A Black Hole's Bottomless Well
When matter becomes extremely dense, it creates a "well" in spacetime (pages 14-15). *Eventually the warping is so severe that anything that strays too close to the resulting hole—matter or even light—falls in and can't escape.*

infinitely dense point known as a singularity.

The term "black hole" was coined in 1967 by U.S. physicist John Archibald Wheeler to describe the region of "no return" that surrounds a singularity. This region, called the event horizon or Schwarzschild radius (after the German theorist Karl Schwarzschild), represents the point beyond which nothing can escape the pull of the black hole's gravity—not even light.

Black holes are not a myth. They exist and can be detected in two ways. When a black hole sucks in dust, gas, and stars, its powerful gravity shreds the material into its atomic components. These atomic particles spiral in, faster and faster, until they reach velocities of over six hundred miles (965 km) per second. As they head toward oblivion, they collide with each other in a heated frenzy, reaching temperatures of several million degrees, at which point X-rays and gamma rays are radiated.

Only black holes are capable of accelerating matter to such high velocities in so tight an orbit; and only black holes are capable of generating X-rays and gamma rays in this manner.

Once the material, and radiation, plunges over the edge, beyond the event horizon, it disappears forever. Where it goes is anyone's guess. At the center of the black hole, where the singularity lurks, the known laws of physics break down.

Singularities are not part of our universe and are not measurable. Since science only deals with measurable quantities, not even science has fathomed the depths of black holes.

The Weight of the Matter
Neutron stars are roughly the size of a small asteroid and yet they have more mass than the Sun. This results in a very powerful gravitational field. A 175-pound (80 kg) person on a neutron star would weigh the equivalent of more than one million tons.

Starlight's Telltale Message

We may never reach the stars—only a couple of space probes have ever left our own solar system, let alone traveled the few trillion miles to our nearest stellar neighbor—yet we know a great deal about the universe around us. What we do know, we have learned largely by decoding the light sent towards us by these distant objects, a study known as spectroscopy.

To understand how the rainbow of colors emitted by an object—its spectrum—can reveal the composition of a star and other distinguishing traits, we need to look at how atoms work.

Atoms are composed of a nucleus, containing positively charged protons and electrically neutral neutrons, and a cloud of negative electrons orbiting the nucleus. The rules of quantum mechanics—the set of laws that govern the behavior of matter at the atomic scale—allow those electrons to occupy only certain orbits with precise energy levels. The transition of electrons between different orbits, or states, can only occur by the absorption or emission of a specific amount of energy. That energy is absorbed or emitted in the form of a packet, or quantum, of light called a photon.

Because the color of light is related to the energy of the photon, each different atomic transition results in a unique color. Many of these transitions have been cataloged by analyzing elements found on Earth in laboratories. Those results can then be used to search for identical results in starlight, roughly the technique that detectives would apply when trying to match a fingerprint on file with those of subjects who they are investigating.

A FULL RANGE OF COLORS

When matter emitting light is very dense and hot, such as star's interior, details of the individual atomic transitions are lost and the material emits a continuous spectrum of colors ranging from red to violet *(below)*. Details of this spectrum depend on only one quantity, temperature of the material, and not on chemical composition or shape. This is the reason that embers in a fire or a poker placed in it—be it iron, steel, copper, or brass—all glow with the same color. The color is an indication of the temperature, varying from relatively cool red to hotter orange and white-hot bluish white. With stars, these telltale colors are sometimes visible to the naked eye. Compare, for

Three Types of Spectrums
Most hot objects, such as stars, emit a continuous spectrum, a band of light that includes all wavelengths, from red to violet (top). *This spectrum will vary depending on the temperature of the body that created it. Individual elements also have their own spectrum. When hydrogen gas is heated, for example, it gives off only a few narrow colored lines that make up the unique spectral fingerprint of that element. This is known as an emission spectrum* (center). *If, however, light from a star passes through a cool cloud of the same gas, the element will absorb from the light the same colored bands produced by an emission spectrum, leaving black lines in their place. The result is an absorption spectrum* (bottom).

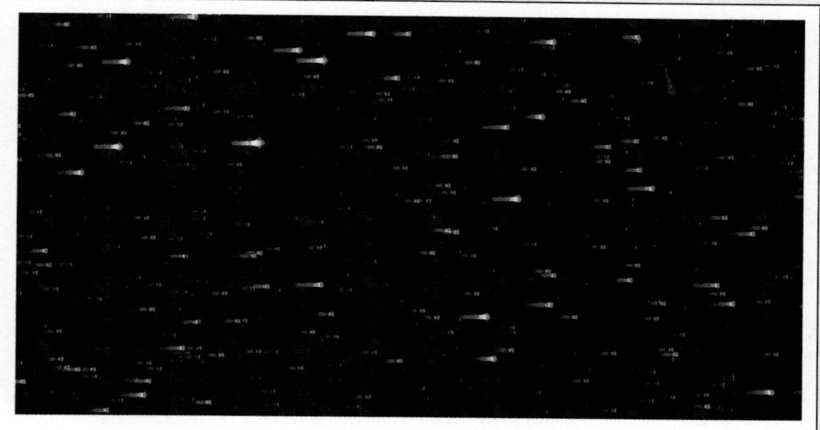

Using a device called a spectrograph, the light from stars can be broken down into a spectrum of colors that reveals much about their composition.

example, the star Antares in the constellation Scorpius with Sirius, the brightest star in the sky, located in Canis Major. The former is reddish, indicating lower temperature, while the latter has the bluish white signature of a hot star.

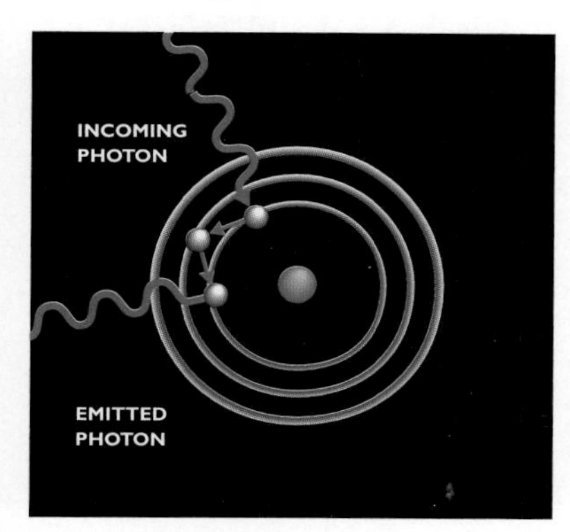

A Question of Energy Levels
Absorption and emission spectrums are the result of the same atomic processes. An incoming photon causes an electron to jump to a higher energy level. When the electron returns to its original orbit, a photon is then emitted from the atom. The wavelength of the photon is equal to the energy difference between the two levels. Since these energy levels are unique for each element, scientists can examine the resulting absorption or emission lines in the spectrum and determine what elements are found in the object.

TWO OTHER SPECTRUMS

In a low-density hot gas, such as the gas of a star's outer atmosphere, the collision between atoms can knock an electron to a higher orbit. When the electron returns to its original orbit, a photon is emitted that corresponds to the difference in energy between the two levels. The result is an emission spectrum *(opposite)*, with specific frequencies, or colors. The pattern and location of the lines in a spectrum are unique for each chemical element.

On the other hand, when the light emitted by the star's hot surface passes through the cooler and less dense outer atmosphere, the photons are removed from the spectrum, creating black "absorption lines." As the light emanating from the star passes through the cool gas, the intervening elements extract from the light the same color bands it would send out if it were heated to a high temperature. The result is that the emission and absorption line patterns are identical *(opposite)*.

THE SOLAR SYSTEM

Forged from a nebula that collapsed under its own gravity, the solar system contains a wide array of celestial bodies— none more remarkable than life-bearing Earth.

> *"To me every hour of the light and dark is a miracle, Every cubic inch of space a miracle"*
>
> — Walt Whitman
> *Leaves of Grass*

Our solar system is the result of a great clash between two opposing forces: the powerful pull of gravity and a counterbalancing centrifugal force. The struggle began some 4.6 billion years ago, when a slowly rotating disklike cloud of interstellar gas and dust, called the solar nebula, gathered in an outer region of the Milky Way galaxy. The gravitational pull of the collected cosmic matter began to collapse the entire structure in upon itself.

Matter within the nebula contracted, then heated up to form a protostar, the precursor to our Sun. At the same time, the nebula's rotation began to speed up, just as skaters spin faster when they pull in their arms. Bits of cosmic matter collided and fused. This process of accretion led to the formation of so-called planetesimals and proto-

The early solar system
As far back as 1755, Prussian philosopher Immanuel Kant suggested that the solar system was formed from a nebula. Kant was correct in his hypothesis that a flat, rotating cloud of interstellar gas and dust was the foundation on which the solar system was built.

We Are Not Alone

Although there are two hundred billion stars in our galaxy, the question of whether there are other solar systems long remained a point of contention among scientists. That all changed in 1995 when Swiss astronomers Didier Queloz and Michael Mayor discovered a planet orbiting a star in the Pegasus constellation. Less than a year later, two more planets were located, one orbiting a star in Ursa Major. Today, thanks in part to the far-reaching eye of the Hubble telescope, some eighteen other planets have been identified in our galaxy, bringing the grand total to twenty-one. So far, none of these extrasolar planets appears capable of supporting life. Does life exist elsewhere in the universe? That is one of the questions astronomers are trying to answer.

planets, the forerunners to the planets, moons, and asteroids that now make up the solar system. The gravitational fields of those bodies helped speed up the process of accretion.

Back at the core of the solar nebula, the temperature of the protostar rose rapidly as matter continued to be drawn inward. That heat eventually caused nuclear fusion, and the star called the Sun was born.

COSMIC INVENTORY

Today the solar system is home to one star, nine planets, sixty-three known moons, an estimated one million asteroids, and countless comets. This collection of celestial bodies is contained within a space that is small by cosmic standards, but still enormous. Sunlight takes just eight minutes to reach Earth, but more than four hours to travel the 3.67 billion miles (5,900,000,000 km) to Pluto.

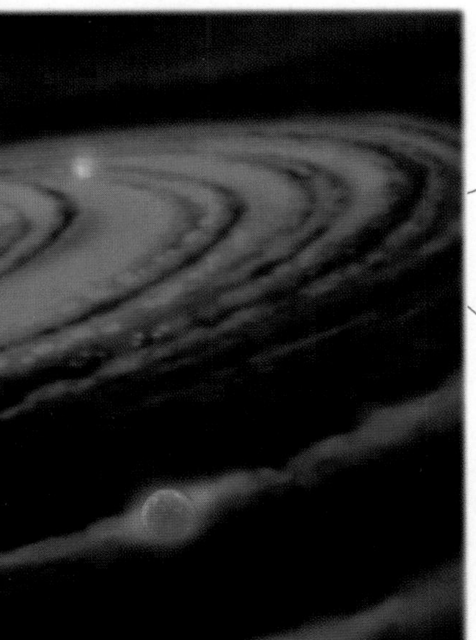

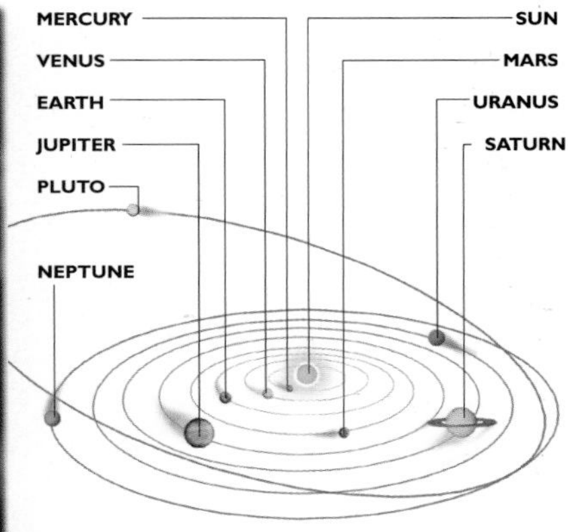

Similar orbital patterns
The planets revolve around the Sun counterclockwise in elliptical orbits. While the solar system is essentially flat—with all the planets situated on roughly the same plane, or ecliptic, as Earth—Pluto's orbit is tilted at seventeen degrees, an indication, some scientists say, that it was either an escaped moon from another planet or a piece of debris left over when the solar system formed.

The Sun

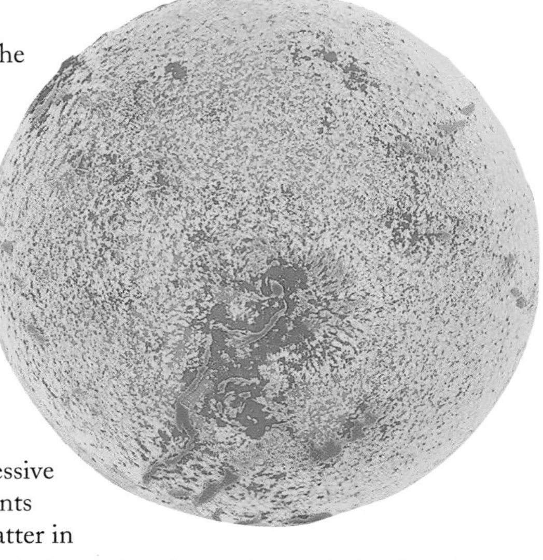

This false-color image of the Sun shows the pockets of cool, dark gas called sunspots (green dots). The grainy, mottled appearance is the result of looking at the top of the convection cells that transport energy to the Sun's surface. Hot, bright, rising gas contrasts with cooler, darker, sinking gas, creating a granulated look.

Although it is the center of the solar system and source of energy for life on Earth, the Sun is nothing special by the universe's standards. Among the two hundred billion other stars that make up our Milky Way galaxy, the Sun is an average-sized stellar body, roughly midway through its term of existence.

Within the solar system, however, the Sun is an impressive and dominant body. It accounts for 98.8 percent of all the matter in the solar system. And its diameter of 865,000 miles (1,400,000 km) is more than a hundred times that of Earth.

The Sun is primarily composed of incandescent gas: about 75 percent hydrogen and 25 percent helium with a trace (0.1%) of heavier elements such as oxygen and nitrogen. Like other stars *(page 24)*, the Sun's core operates as a massive thermonuclear reactor. There,

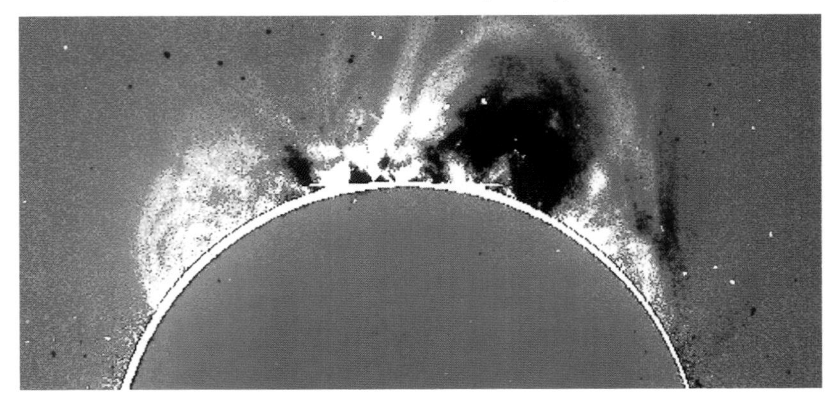

A solar flare rises into the Sun's outer atmosphere: the corona. Caused by bursts of magnetic activity in the Sun, the bright plumes of gas stream away from the Sun to form part of what is called the solar wind: protons and electrons that travel through space at up to six hundred miles (970 km) per second.

where the temperature tops twenty-seven million degrees Fahrenheit (15,000,000°C), hydrogen atoms collide and form helium. The transformation results in the loss of five million tons of matter a second—and the release of energy in the form of particles called photons that take upward of one million years to move from the Sun's center to its surface.

THE PHOTONS' JOURNEY

The first leg of the photons' journey carries them through the radiative zone, where they collide with atomic particles losing energy along the way. Next, in the convection zone, the photons are absorbed by hot gases, which bubble toward the surface in the form of giant cells. By the time they reach the convection zone's outer edge, the photons have cooled to ten thousand degrees Fahrenheit (5,500°C).

The Sun's surface, called the photosphere, seethes with huge convection cells called granules *(opposite, top)*. Some thousand miles (1,600 km) in diameter, each granule is a pocket of hot gas rising from below. It is during the continual process of heating and cooling that the bubbling granules release their radiant energy into space—equivalent to a hundred billion one-megaton nuclear bombs exploding a second.

THE FINAL DAYS

The Sun will continue to produce energy at this rate for approximately another five billion years before its supply of hydrogen runs out. As its surface temperature drops substantially, the Sun will begin its transformation from a star to a so-called red giant, a bloated, cooler relic of its former self. It will eventually swell out further than Venus's orbit.

After one billion years in this state, the Sun will shed the remnants of its gaseous atmosphere and the remaining matter will collapse on itself forming a highly compressed core that is so dense, a single teaspoon would weight more than five tons. This glowing white-hot core, called a white dwarf, will gradually radiate all its energy until it becomes a dark, burnt-out cinder.

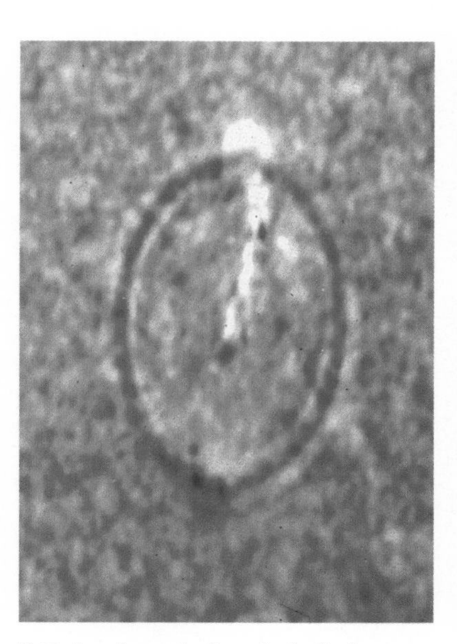

Explosions from solar flares in the Sun's atmosphere send shock waves to the surface that result in massive sunquakes, shown in this photo by the Solar and Heliospheric Observatory (SOHO). This 1996 sunquake contained about forty thousand times the energy released in the great San Francisco earthquake of 1906.

The Terrestrial Planets

On a tour of the solar system that starts from the Sun and moves outward, the first four stops would be Mercury, Venus, Earth, and Mars. These are known as the terrestrial planets. They are rocky in nature, with some similar physical and chemical properties to Earth—vastly different from the gassy giants in the outer solar system *(page 44)*.

FIRST STOP: MERCURY

Located a mere thirty-six million miles (58,000,000 km) from the sun, Mercury is the innermost planet in the solar system. It is also the second smallest, exceeding the size of only Pluto. Because it is situated on the Sun's doorstep, Mercury's surface temperatures soar to eight hundred degrees Fahrenheit (427°C). The planet's nighttime temperature, though, can plummet to minus 280 degrees Fahrenheit (-173°C), the result of Mercury's unique orbital pattern. While Mercury takes only eighty-eight days to revolve around the Sun, it spins on its axis only once every fifty-nine days. As a result of this ratio, one face of the planet can lie in total darkness for weeks.

Ironically, Mercury's proximity to the Sun makes it one of the darkest planets in the solar system, as seen from Earth. The extreme heat has made it impossible for the planet to maintain an atmosphere remotely as dense as its two nearest neighbors, Venus and Earth. Consisting mainly of captured gases from the solar wind, Mercury's atmosphere is 10 to 12 percent as dense as that of Earth's. Surface rocks reflect a mere 8 percent of the light they receive.

In the mid-1970s, the space probe *Mariner 10* took photographs that revealed much about Mercury's history. Deep craters, including one the size of Texas, attest to the constant barrage the planet has endured from meteorites. But the images also showed huge plains on the surface. Scientists speculate that Mercury was once liquid rock and cooled over time. While bombardment from small meteorites lightly cratered Mercury's surface, larger ones pierced its crust and loosed torrents of lava, which spread out in plains. The crust is crisscrossed by long, wavy scarps. These clifflike struc-

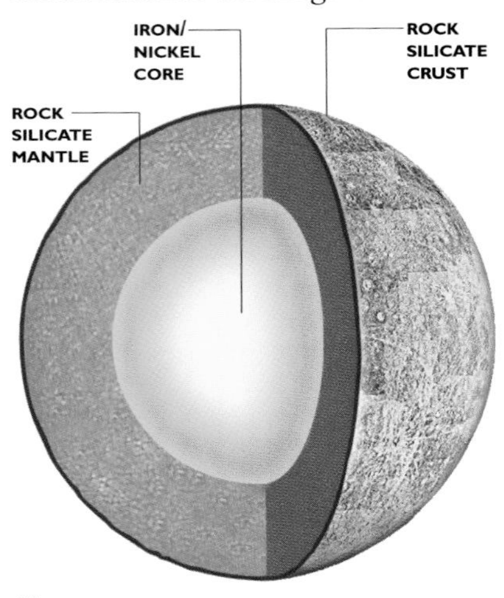

Mercury
Diameter: 3,029 miles (4,875 km)
Rotation: 59 days
Orbit: 88 days
Number of moons: 0
Distance from sun: 36,000,000 miles (58,000,000 km)

tures, which range up to ten thousand feet (3,048 m) in height, formed when the planet's diameter shrank by roughly a mile (1.6 km) as it cooled. The planet's core is composed of iron and nickel. In fact, Mercury is richer in iron, by percentage, than any other planet in the solar system.

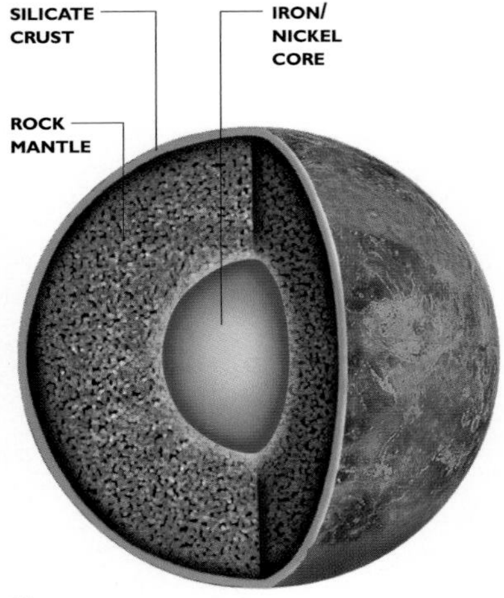

Venus
Diameter: 7,521 miles (12,104 km)
Rotation: 243 days
Orbit: 243 days
Number of moons: 0
Distance from sun: 67,000,000 miles (108,000,000 km)

In this radar image of Venus, taken by the orbiting Magellan probe, blue represents the lower elevations; green, brown, pink, and white the higher. Near the top of the picture lies the upland continent called Ishtar Terra.

VISITING VENUS

Unlike its virtually atmosphereless neighbor, Venus is cloaked in dense, reflective clouds composed mostly of sulfuric acid so thick that only 2 percent of the Sun's rays penetrates it to reach the planet's surface. However, as the surface is warmed, it emits infrared radiation that is trapped by the carbon dioxide in the atmosphere above. The result: a devastating greenhouse effect in which heat never escapes, leaving Venus with a metal-melting surface temperature of 850 degrees Fahrenheit (454°C).

Unable to penetrate Venus' layered cloud cover with standard telescopes, scientists have gathered valuable information about the planet's surface by means of a series of space probes beginning in 1962. The most recent craft, *Magellan*, was launched in 1989 and mapped the majority of the planet by bouncing radar waves off the Venusian surface.

Scientists once believed that Venus had large bodies of water and possibly even supported plant life. But the images gleaned by probes have revealed a barren and inhospitable topography. Massive mountain ranges and towering volcanoes jut up from the rocky landscape, and relatively young lava plains indicate that some volcanoes still may be active.

In fact, some scientists believe that volcanism largely resurfaced Venus in the last five hundred million years or so. This could explain why Venus has relatively few craters. Other astronomers suggest that this lack could be traced to the planet's thick atmosphere, which acts as a protective shell and burns up smaller meteorites before they reach the surface.

EARTH: THE HOME PLANET

Life is the one quality that separates Earth from all the other planets in our solar system. Earth teems with it—roughly 1.75 million species have been classified, with thousands of new ones being added to the list each year. In fact, zoologists believe that we have scarcely scratched the surface when it comes to discovering new life forms. They estimate that there could be as many as fourteen million species on the planet.

The fact that life can exist at all on Earth is testament to conditions unique in the solar system, beginning with a multi-layered atmosphere, which filters the Sun's deadly radiation while still permitting sunlight to reach the planet's surface. As the Sun's rays pass through the atmosphere, lethal gamma and X-rays are absorbed in the uppermost level, called the thermosphere. Ozone in the stratosphere keeps most of the dangerous ultraviolet rays from striking Earth's surface. The atmosphere's lowest level, the troposphere, soaks up much of the Sun's infrared rays.

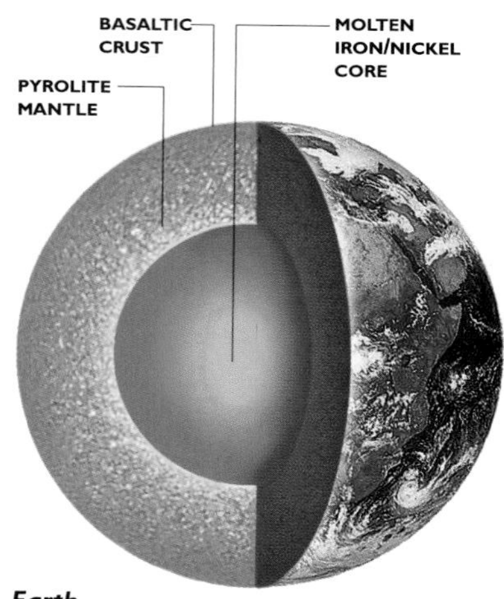

Earth
Diameter: 7,918 miles (12,743 km)
Rotation: 24 hours
Orbit: 365.25 days
Number of moons: 1
Distance from sun: 93,000,000 miles (150,000,000 km)

A Moonless Earth

The Earth without the Moon? It's hard to imagine. But the difference would be more than just the lack of a celestial object to inspire poets and a familiar sight in the night sky.

Speculation about conditions on a moonless Earth is based largely on a subtraction of the Moon's effect on the tides. The complex tidal interplay of Earth and Moon produces a drag on Earth's rotation, slowing it by two thousandths of a second per century.

Three billion years ago, the Earth completed a single rotation in just eight hours. This rapid spinning would have transferred energy to the atmosphere and oceans, producing high winds and fast ocean currents, while the Moon's proximity—sixty thousand miles (96,600 km) closer at the time—would have produced greater tidal surges. In turn, these mixing effects would have most likely accelerated the appearance of life by churning the primordial brew of air and sea.

Without the Moon's stabilizing effect, the Earth might still be spinning faster; raging hurricanes would probably prevail, and life would certainly not exist as we know it.

From the moon, Earth appears as a huge blue globe partially veiled by clouds. Some 71 percent of the Earth's surface is covered by water. This photo was taken by the Apollo 11 *astronauts in July 1969.*

Made up of nitrogen, oxygen, argon and carbon dioxide, Earth's atmosphere also helps keep the planet at the ideal temperature for liquid water—another essential building block for life.

Like its atmosphere, Earth itself is made up of layers. The planet's center is composed of a solid iron-and-nickel core surrounded by an outer core of molten metal that scientists believe can reach temperatures in excess of nine thousand degrees Fahrenheit (5,000°C).

The mantle, the largest section of the inner planet, is pyrolite, a rock rich in iron and magnesium. Riding atop this churning mass is Earth's thin crust, made up of a dozen segments, or plates, that continually shift and collide. Volcanic eruptions and earthquakes occur along plate boundaries as the massive, mobile segments grate against or slide under each other, ceaselessly changing Earth's face.

THE MOON: EARTH'S PARTNER

Literally a chip off the old block, the Moon is believed to have been formed approximately 4.6 billion years ago when Earth was hit by a Mars-sized body. The collision fired debris—mostly crust and mantle—into orbit. Over time (some scientists believe in as little as a few thousand years), the material coalesced to create the Moon.

Ironically, the Moon has little in common with the planet from which it was born. Unlike Earth, the Moon possesses no liquid water, weather, or life forms. And while Earth is in a

Pocked with craters, the far side of the Moon displays the scars of a series of violent meteor showers that ended about 3.9 billion years ago.

Barren and yet eerily familiar, the Martian landscape was captured by the camera of Mars Pathfinder *during its 1997 mission to the Red Planet. In the foreground is the ramp that allowed a robotic rover, called* Sojourner, *to drive down to the Martian surface.* Sojourner, *which weighed twenty pounds (9 kg) and was operated by solar panels, can be seen at right as it investigates a rock nicknamed Yogi. The circular tracks near the ramp were the result of an experiment the rover made to analyze the Martian soil.*

state of constant geological activity, the Moon lies dormant, its cratered surface shaped instead mostly by the onslaught of meteorites. However, recent discoveries have shown that the Earth and the Moon share one common trait. In the winter of 1998, the *Lunar Prospector* spacecraft discovered ice on the Moon's surface at the lunar poles.

Because the Moon is so close—just 240,000 miles (386,000 km)—its larger surface features can be clearly seen from Earth using nothing more than binoculars *(pages 98 to 99)*. But stargazers always see the same lunar face. This is because the Moon takes exactly as long (twenty-seven days and seven hours) to rotate on its axis as it does to orbit Earth. Scientists believe that the Moon used to rotate more quickly than this, but that the Earth's gravitational influence has slowed it down.

Mars' Grand Canyon

The mammoth Valles Marineris extends for more than twenty-five hundred miles (4,000 km), one-fifth of Mars' diameter. First detected by *Mariner 9* in 1971, the huge rift could easily accommodate the Grand Canyon in one of its smaller offshoot canyons.

MARS: COLD, RED DESERT

As well as being next-door neighbors in the solar system, Mars and Earth share many of the same characteristics. Both planets are orbited by satellites, though Mars' two, Phobos and Deimos, are tiny compared with Earth's moon—each less than fifty miles (80 km) in diameter. Shifting sand dunes, clouds, storms, and polar ice caps are common to both worlds. Similar to Earth, Mars takes twenty-four hours and thirty-seven minutes to complete a full rotation on its axis—an axis tilted at an Earthlike twenty-five degrees to its orbit.

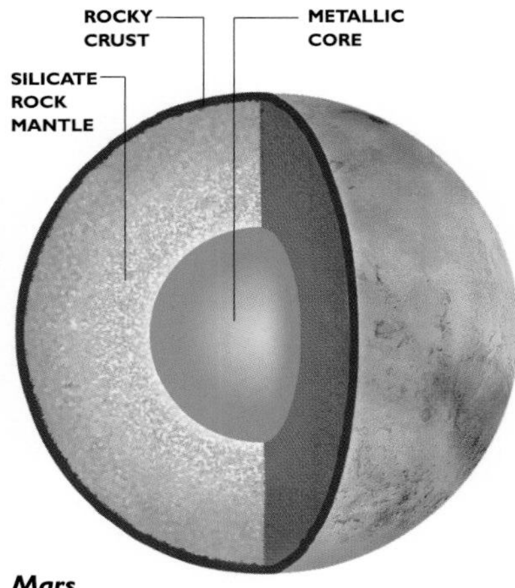

Mars
Diameter: 4,213 miles (6,780 km)
Rotation: 24 hours, 37 minutes
Orbit: 687 days
Number of moons: 2
Distance from sun: 141,700,000 miles (228,000,000 km)

While both planets enjoy four distinct seasons, that's where the climactic similarities end. Mars is a barren world with an atmosphere made up of 95 percent carbon dioxide. Unlike Venus, on which thick layers of carbon dioxide create blistering surface temperatures, Mars' thin atmosphere—less than 1 percent as dense as Earth's—retains little heat. Surface temperatures rarely exceed the freezing point of water and at night can dip as low as minus 190 degrees Fahrenheit (-123°C).

Mars is called the Red Planet because of its rusty hue. Huge deserts of red oxidized sand cover much of the surface and are punctuated by ocher-colored boulder fields and lava flows. Violent wind storms can create dust clouds that have obscured the planet's surface for weeks at a time.

Water exists on Mars in its frozen polar caps and in its atmosphere. Scientists speculate that the huge channels visible on the surface may have been gouged by raging floods during a watery phase of the planet's distant past. This, and possible volcanic activity, may have carved the Valles Marineris, an enormous chasm that stretches some twenty-five hundred miles (4,000 km).

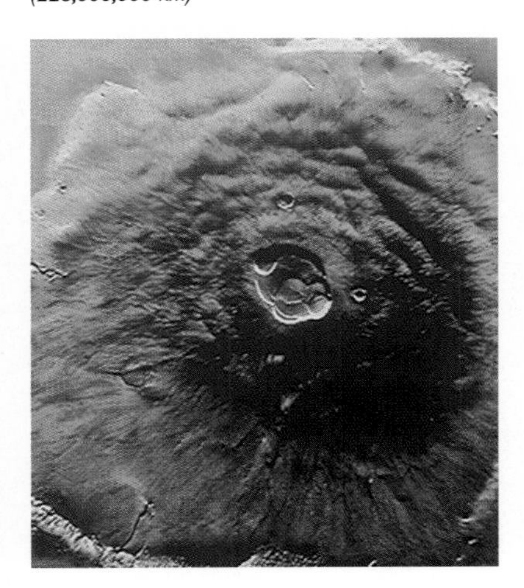

A dormant Martian volcano, Olympus Mons stands some seventeen miles (27 km) high—almost three times the height of Mount Everest—and is thought to be the tallest peak in the solar system.

Probing the Solar System

If all goes as planned, NASA's *Cassini* mission will arrive at Saturn on July 1, 2004, six years and eight months after its launch from Earth on a 2.2-billion-mile (3,500,000,000-km) odyssey. For the next four years, a battery of scientific equipment aboard the twenty-two-foot-(7-m)-long craft will scrutinize the planet and its moons, analyzing their chemical composition, investigating Saturn's magnetic field, radar mapping the surface features of its largest moon, Titan, and beaming back thousands of images taken by a camera so sensitive it could see a quarter on the ground from 2.4 miles (3.9 km) in the air. Meanwhile, a probe called *Huygens* will descend from *Cassini* to Titan by parachute, using an onboard robotic laboratory to explore the clouds, atmosphere, and surface of the planet-sized satellite.

Scientists have designed *Huygens* to function even if it lands in the oceans of ethane and methane that many astronomers believe cover Titan's surface. The information *Huygens* gathers will be relayed up to *Cassini* orbiting above, which will then transmit the data back to Earth, an approximately eighty-minute trip at the speed of light.

Scientists expect that the information from *Cassini* and *Huygens* will revolutionize our understanding of one of the solar system's most enigmatic planets and its moons, providing insights that would elude even the far-seeing eye of the Hubble space telescope.

INTREPID EXPLORERS

Space probes have come a long way since the early years of space travel, when missions like the 1964 *Ranger 7* flight were designed to crash land into the moon, beaming back fuzzy photos until the moment of impact. Now probes are capable of reaching the farthest parts of the solar system—and beyond. Some are even designed for return trips. For example, the *Stardust* mission, launched in February 1999, will fly through the tail of Comet Wild in January 2004, then return to Earth with

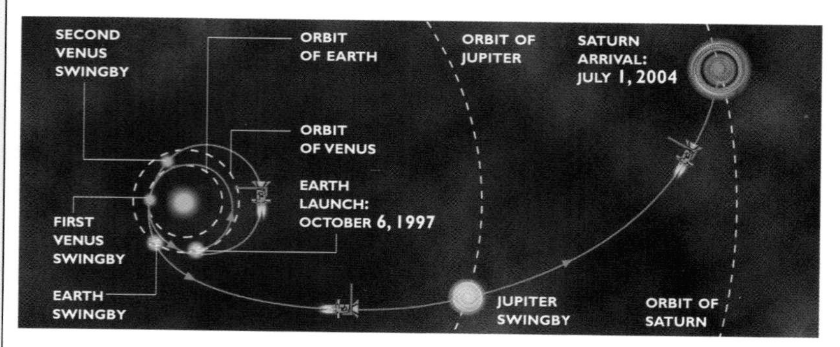

"Slingshotting" its Way to Fuel Efficiency

After swinging by Venus twice on its route to Saturn, the Cassini probe will pass by Earth (August 18, 1999) and Jupiter (December 30, 2000), using the gravity of the planets to boost the craft along to its target. The circuitous route will provide the energy equivalent to seventy-five tons of rocket fuel.

samples of interstellar dust particles two years later.

Space probes generally fall into one of three categories: flyby, orbiter, and lander. In a flyby mission, such as *Voyager*'s Grand Tour of the solar system in the 1970s and 1980s, the craft makes a one-time pass by a planet. Orbiters, such as *Cassini*, station themselves in orbit, allowing long-term analysis of a planet below. Landers attempt to set down on a planet's surface. Different landings call for different techniques. While *Huygens* will rely on parachutes, the *Mars Pathfinder* mission employed air bags to cushion its landing, bouncing like a giant ball.

With the increasing distances that spacecrafts travel, NASA engineers are continually faced with new challenges. When the *Voyager 2* craft passed by Neptune in 1989, for example, scientists had to deal with sunlight nine hundred times fainter than on Earth. How to take a photo requiring a thirty-second exposure while the spacecraft sped by the planet at fifty-five thousand miles (88,514 km) per hour? Engineers solved the problem by sending *Voyager* a message instructing the entire craft to pan while the time-exposure shots were taken. The result: razor-sharp images of a planet almost three billion miles (5,000,000,000 km) away.

Then there is the challenge of keeping the craft on course. Engineers employ various methods to keep track of spacecrafts, sometimes years after they are launched. One tactic is ranging: sending signals to the spacecraft and measuring the time it takes to receive a reply. Knowing the speed of light, engineers can determine the exact distance the craft is from Earth. Another technique relies on optical data, examining photos beamed back from the craft and examining the target body or one of its moons against a known star background to know where the spaceship is with respect to its destination.

Closing in on Saturn
In 2004, the Cassini *spacecraft will fire its rocket to slip into an orbit around Saturn and explore the planet.*

Mid-course corrections can be made by firing thrusters on the probe. As one NASA engineer says, it's a lot easier to sink a billiard ball on a long shot if you can nudge it with the cue stick a couple of times on its way to the pocket—especially if the pocket is more than a billion miles (1,600,000,000 km) away.

The Outer Planets

The solar system's five outer planets can be divided into two categories: the Jovian planets and Pluto. Unlike the four terrestrial planets, Jupiter, Saturn, Uranus, and Neptune have no solid surface. Instead, these massive planets are composed primarily of gases such as hydrogen and helium. Situated on the very outskirts of the solar system, the rock and ice world of Pluto stands alone—a seeming anomaly that excites debate among astronomers.

JUPITER: PLANETARY GIANT

Weighing more than the rest of the other eight planets combined, Jupiter is by far the largest planet in the solar system. With no solid surface, Jupiter is blanketed by a six-hundred-mile (966-km)-deep cloud cover that can be seen with a telescope as colorful bands that encircle the planet. As gas is heated by Jupiter's seething interior of hydrogen and helium, it rises to create light-colored bands, called zones. The darker swirls, called belts, are gases that are cooling and sinking.

Of Jupiter's sixteen known moons, Europa is one of the most intriguing. Scientists had long thought the satellite to be a dead and frozen world. However, photographs taken by the *Galileo* spacecraft in 1998 show a remarkably crater-free surface, leading astronomers to suggest that a sub-

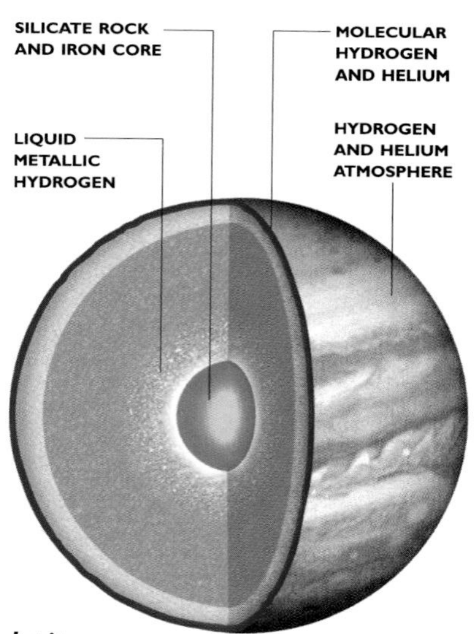

Jupiter
Diameter: 88,846 miles (142,984 km)
Rotation: 9.925 hours
Orbit: 11.8 years
Number of moons: 16
Distance from sun: 483,600,000 miles (778,300,000 km)

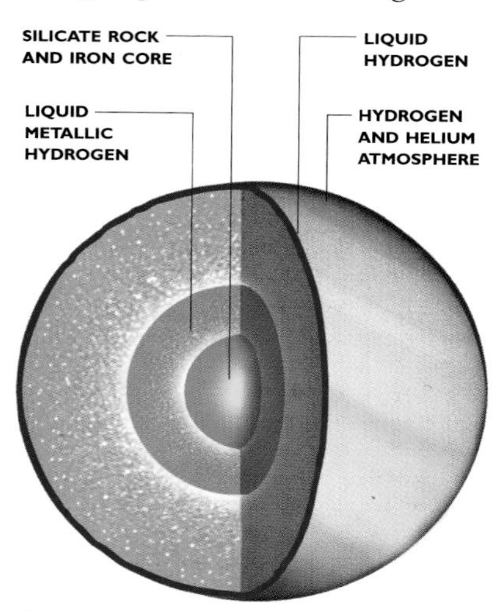

Saturn
Diameter: 74,898 miles (120,537 km)
Rotation: 10.6 hours
Orbit: 29.4 years
Number of moons: 18
Distance from sun: 889,800,000 miles (1,432,000,000 km)

surface ocean has smoothed out impact scars on the icy surface over the last few million years.

SATURN: RINGED WONDER

Like its neighbor Jupiter, Saturn is a sea of liquid hydrogen and helium, cloaked in thick clouds. Composed for the most part of crystallized ammonia, these clouds are whipped along by ferocious winds that can exceed a thousand miles (1,600 km) per hour. But Saturn's most prominent characteristic is the ring system girding its equator. While the other Jovian planets also have rings, Saturn's are easily the most pronounced and are visible from Earth using just a small telescope.

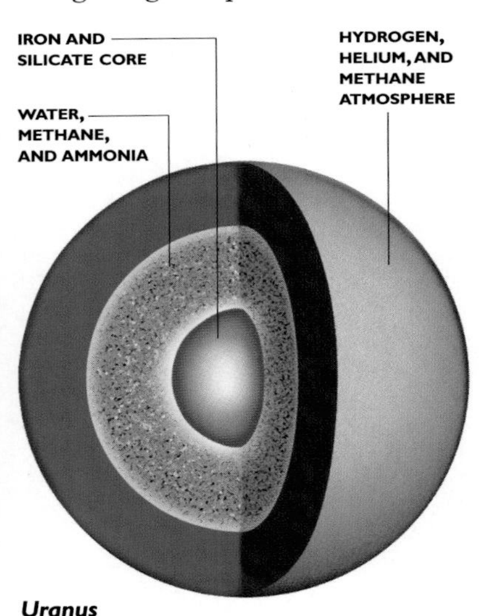

Uranus
Diameter: 31,763 miles (51,118 km)
Rotation: 17.24 hours
Orbit: 84.1 years
Number of moons: 17
Distance from sun: 1,784,000,000 miles (2,871,000,000 km)

Divided into seven zones and spanning some 170,000 miles (274,000 km) in diameter, Saturn's rings are made up primarily of ice, with some dust and rock mixed in. Scientists suggest that this debris is the remnants of an ancient moon disrupted by tidal forces.

Eighteen moons are known to orbit Saturn. In 1995 the Hubble telescope detected four other bodies that may add to that total.

Of particular interest to scientists is Titan, the largest of Saturn's satellites. Titan is the only moon in the entire solar system to possess a dense atmosphere. Composed largely of nitrogen and methane, Titan's atmosphere is also believed to contain hydrogen cyanide and cyanogen—organic compounds that were the building blocks in the evolution of life of the Earth.

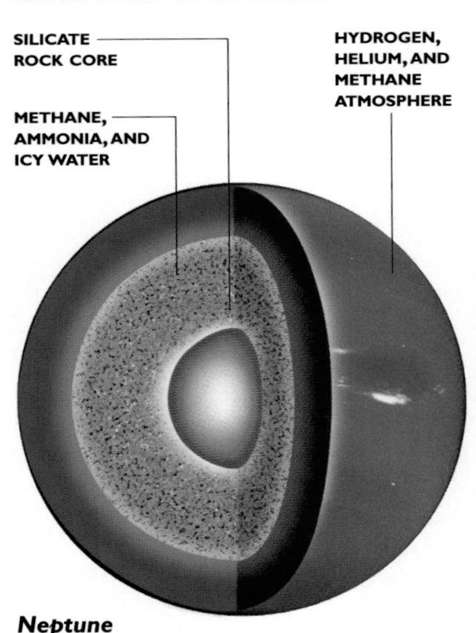

Neptune
Diameter: 30,775 miles (49,528 km)
Rotation: 16.1 hours
Orbit: 164.9 years
Number of moons: 8
Distance from sun: 2,795,000,000 miles (4,498,000,000 km)

URANUS: TOPPLED PLANET

Situated almost nine hundred million miles (1,500,000,000 km) from Saturn, Uranus wasn't discovered until 1781, when British astronomer William Herschel spotted it. The finding immediately doubled what had been thought to be the size of the solar system.

Uranus remained a relatively unknown planet until 1986, when the *Voyager 2* probe passed by the planet. Chief among the probe's findings were ten new moons and a pair of new rings.

Armed with instruments able to penetrate the dense clouds, the probe found that the planet's average temperature of minus 350 degrees Fahrenheit (-212°C) varied very little between the equator and the poles. This is because, unlike the other planets in the solar system, Uranus' axis of rotation lies almost in the plane of its orbit rather than roughly perpendicular. Scientists speculate that a violent collision with a large body may have jarred the planet onto its present-day axis, ejecting gas and debris that coalesced to form the seventeen moons that orbit the planet.

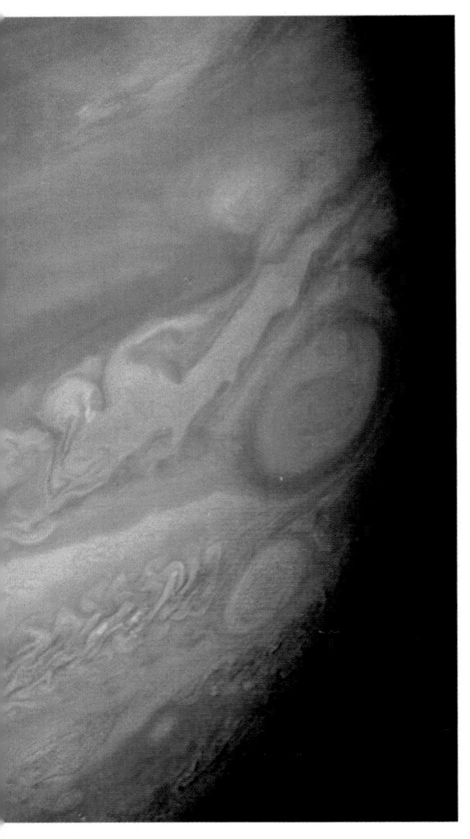

Because Jupiter rotates at such a rapid rate, its cloud patterns are constantly changing. One notable exception is the Great Red Spot, a high-altitude storm cloud that was first observed in the 1600s.

NEPTUNE: TURBULENT WORLD

The most distant of the Jovian planets, Neptune is also one of the most violent. The planet's methane cloud cover is whipped across its face by winds that can gust up to twelve hundred miles (1,931 km) per hour. Huge spots that dot the planet's blue atmosphere are massive storms. The largest was the Great Dark Spot, first spotted by *Voyager 2* in 1989; bigger than Earth, it has since disappeared.

Beneath Neptune's frigid layer of hydrogen, helium, and methane lies a layer of water, ammonia, and liquid methane that covers its rocky iron core. Of all the Jovian planets, Neptune possesses the least pronounced ring system. Data compiled by *Voyager 2* proved that the rings actually consist of four

This image, taken by the Galileo spacecraft, shows two volcanic eruptions on Jupiter's moon Io. One plume can be seen on the bright limb or edge of the moon. The second eruption, near the center close to the terminator (the boundary between day and night), appears as a gray circle with a red plume extending downward.

incomplete arcs, rather than groupings of particles, that encircle the entire planet.

Voyager 2 also found six of the planet's eight moons. Triton, largest of these, is thought to be the coldest site in the solar system, with temperatures plummeting to minus 391 degrees Fahrenheit (-235°C). Surprisingly active geologically, Triton contains nitrogen geysers that erupt as high as five miles (8 km).

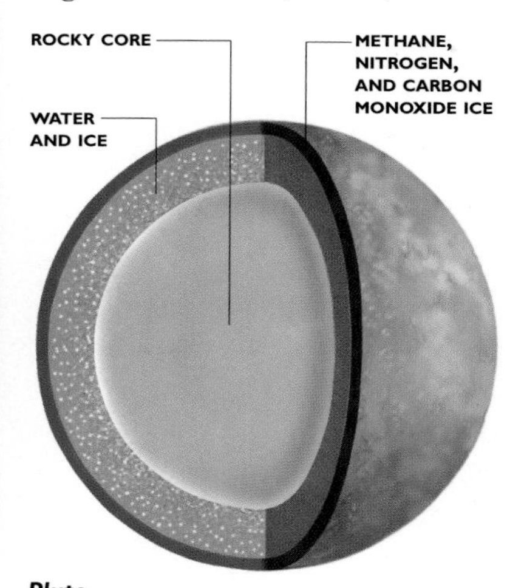

Pluto
Diameter: 1,432 miles (2,305 km)
Rotation: 6.4 days
Orbit: 248.6 years
Number of moons: 1
Distance from sun: 3,675,000,000 miles (5,914,000,000 km)

PLUTO: THE GREAT DEBATE

The smallest and most distant planet in the solar system, Pluto is also at the center of a heated astronomical debate: whether or not it merits classification as a planet.

Just two-thirds the size of Earth's moon, Pluto has a highly elliptical inclined orbit, very different from the more circular paths followed by the other planets. In fact, Pluto's irregular circuit crosses the orbital path of the planet's nearest neighbor, Neptune. This means that when it takes the "inside track," as the planet did between 1979 and 1999, Pluto is actually closer to the sun than to Neptune.

Some astronomers now believe that Pluto is merely one of the larger trans-Neptunian objects (TNOs), icy bodies with orbits that lie beyond—or crisscross—that of Neptune. There are about one hundred TNOs now known to exist.

Interplanetary Bodies

Making its first pass through the inner solar system in forty-two hundred years, Hale-Bopp (above) *came closest to Earth in spring 1997. The dazzling comet has a nucleus twice the size of Halley's Comet.*

Most interplanetary bodies, such as comets, meteors, and asteroids, orbit hundreds of millions—or billions—of miles from Earth. Some maverick bodies, however, make the occasional, often spectacular, foray into the inner solar system.

COMETS: GREAT BALLS OF ICE

Flying in the face of its image as a celestial fireball, a comet is really a "dirty snowball" of water, ice, carbon monoxide, rock, and carbon-rich dust. Comets with relatively brief orbits generally come from a spherical region beyond Neptune called the Kuiper Belt. Long-period comets generally originate within the Oort Cloud, a spherical shell-like reservoir of billions of comets that surrounds the solar system.

As a comet approaches the sun during its elliptical orbit,its nucleus begins to heat up and evaporate. Dust and gas are ejected from its

The asteroid Gaspra 951 measures eleven by seven by six miles (18 by 11 by 10 km).

surface to form a halo, or coma. The pressure of sunlight drives gas and dust from the coma to make the comet's tail—and dazzles skywatchers. While the coma can span some sixty thousand miles (96,560 km), the comet's tail can extend for millions of miles.

Meteors, or shooting stars, are the literal offshoots of comets. Starting as residue from a comet's dust, these small particles burn up when they hit Earth's atmosphere.

ASTEROIDS: MINOR PLANETS

Because they are composed of rock and iron, asteroids are sometimes referred to as minor planets. Most of the estimated one million asteroids—thought to be leftover debris from the creation of the solar system—orbit the sun in a belt between Mars and Jupiter. They range in size from rock-sized pieces to the massive body named Ceres, which measures nearly six hundred miles (965 km) across.

Each day falling debris from space, such as the meteorite above, adds an estimated one hundred to one thousand tons to Earth's overall weight.

Planetary Emissary

Although its name is uninspiring, meteorite ALH 84001 has helped rekindle a long-standing debate: Does life exist on other planets in the solar system?

The potato-sized lump of Martian rock, ejected by a meteor impact long ago, was discovered in 1984 in Antarctica. Twelve years later, scientists announced that the carbonate crystals found in the rock's makeup were likely formed by biological activity. Even more enticing was their suggestion that microscopic clusters on the rock were possibly fossils of long-dead Martian microbes.

However, scientists on the other side of this cosmic pro-life debate maintain that the carbonate crystals were born not of biology, but of violent, high-temperature impacts. Life on Mars? The debate continues.

Though remote, the chance of a large asteroid colliding with Earth is not an impossibility. Many scientists believe that one such cataclysmic event caused the extinction of the dinosaurs sixty-five million years ago. In 1991 an asteroid about thirty feet (9 m) across came within just 106,000 miles (171,000 km) of hitting Earth.

Each day, about ten meteorites—asteroid fragments—pepper the planet's surface. Fifty thousand years ago, a meteorite plowed into the area now known as Arizona. The resulting Barringer crater, measuring 3,960 feet (1,207 m) across and 600 feet (183 m) deep, serves as a warning of the power these wayward travelers possess.

THE CUTTING EDGE OF ASTRONOMY

No longer held back by the limitations of Earth-based optical telescopes, astronomers are relying on cutting-edge equipment to gaze deeply into space and expand our knowledge of the universe.

Although the night sky presents a light show of infinite riches, optical astronomy represents only a small part of the window through which astronomers look to unlock the secrets of the universe. In fact, visible light occupies a narrow band of the range of energy produced by all celestial objects, known as the electromagnetic spectrum.

> *"Astronomy compels the soul to look upward and leads us from this world to another."*
>
> — PLATO
> *The Republic*

Electromagnetic radiation propagates in the form of waves. But depending on how it is measured, it can also exhibit particle-like properties manifested as massless bundles of energy called photons. Each type of radiation is distinguished by its wavelength and its frequency—the number of wavelengths that pass a given point per second.

At one end of the electromagnetic spectrum are low-frequency, long-wavelength radio waves. Following are microwaves, infrared radiation, visible light, ultraviolet radiation, X-rays, and, finally, powerful, high-frequency gamma rays.

Earth is under constant bombardment by all these forms of radiation, but most of it is absorbed by the atmosphere before reaching the ground. Scientists therefore rely on space probes and orbiting observatories to better observe and measure these otherwise absorbed portions of the electromagnetic spectrum.

Radio waves are unaffected by atmospheric absorption, which makes them one of the most useful tools in astronomy. Their penetrating

Stellar processes occur throughout the Milky Way galaxy at different energies, each emitting its own characteristic wavelengths of electromagnetic radiation. By studying different wavelengths, astronomers learn about the Milky Way's structure, its evolution, and the processes taking place within it.

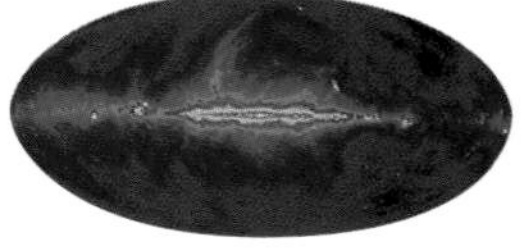

RADIO

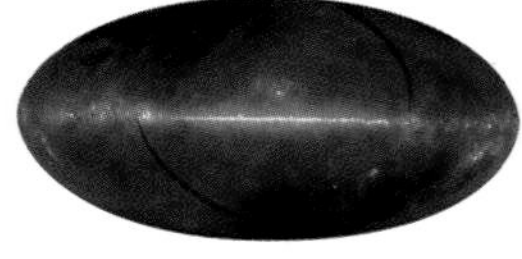

INFRARED

VISUAL

ability allows radio telescopes to capture information coming from dust-obscured regions, such as the center of the Milky Way galaxy. At radar wavelengths, they can penetrate the dense cloud cover of a planet such as Venus, and bounce off its surface to create accurate topographical maps.

Light is the only form of electromagnetic radiation that is visible to the human eye. The colors of the visible spectrum are distinguished by their wavelength—red being the longest, violet the shortest.

Unless blocked by clouds, visible light is able to penetrate the Earth's atmosphere. Still, it is sensitive to the distorting effects of atmospheric turbulence. The Hubble Space Telescope, currently in orbit high above the Earth's filtering atmosphere, ranks as astronomy's most clear-sighted visual observatory.

Gamma rays are produced by sources ranging from solar flares and pulsars to quasars and mysterious gamma-ray bursters—intense, distant blasts of radiation. At the present time, the Compton Gamma-Ray Observatory is investigating these sources from its lofty orbit 270 miles (435 km) above the Earth.

Measuring the Electromagnetic Spectrum: Tools of the Trade

In order to study the entire electromagnetic spectrum scientists use a range of land-based and orbital observatories, including the Far Ultraviolet Spectroscopic Explorer (FUSE), the Chandra X-Ray Observatory, the Compton Gamma-Ray Observatory, the Very Large Array (VLA), the Submillimeter Telescope (SMT), and the Keck Telescopes.

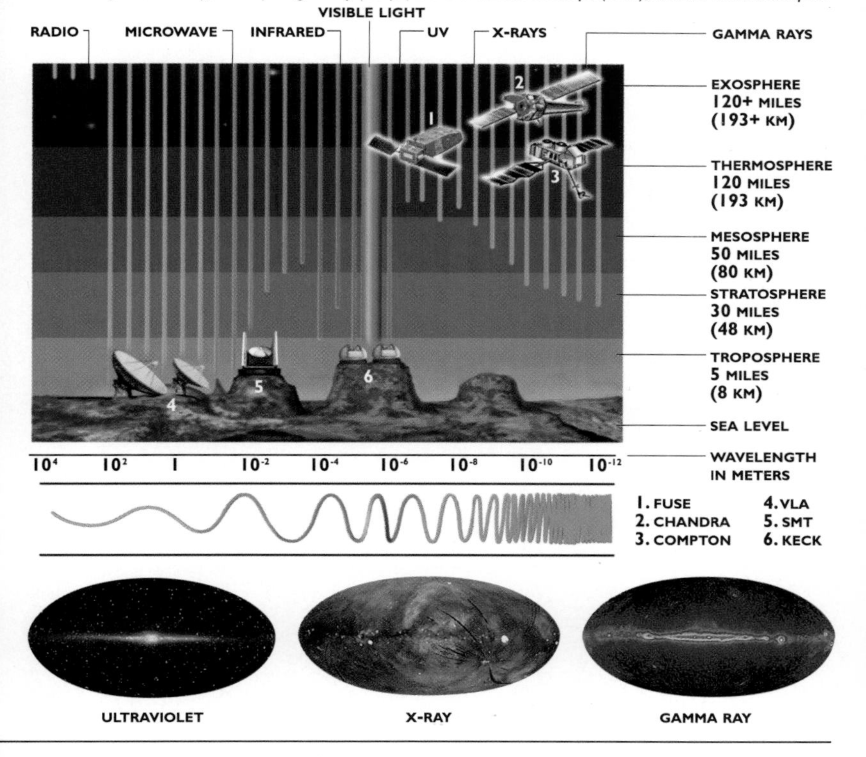

Tuned In

Stretched across the dry plains of San Augustin, New Mexico, twenty-seven dish-shaped radio telescopes operate in unison as they gather in the faint radio emanations of the universe. Since its completion in 1980, the Very Large Array (VLA) has established itself as one of the world's premier observatories.

The super-sensitive VLA antennas pick up radio waves that reach Earth from both near and far. These dishes have recorded radiation from interstellar clouds where stars are forming, detected spurts of high-energy electrons from active galaxies that are powered by black holes, and investigated the atmosphere of the planets in our solar system.

The Very Large Array provides images that are far more than mere complements to those produced by visible light. The center of the Milky Way galaxy, for example, is invisible to optical telescopes because of the thick dust spread across the intervening space. But the VLA has succeeded in drawing back that veil, providing astronomers with new insights into a galactic center that they have now determined is some ten million times the mass of the Sun.

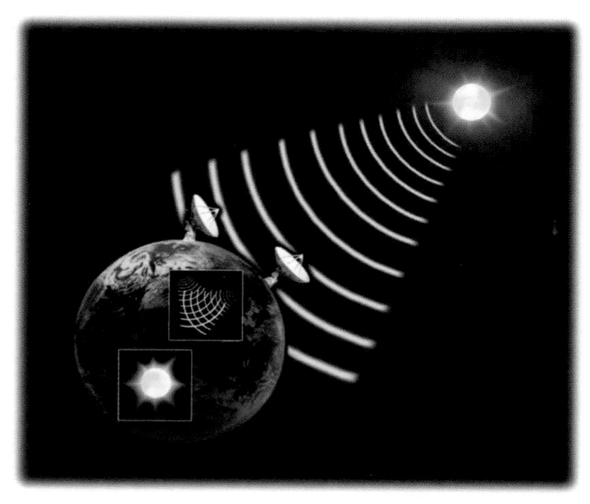

Working in Tandem
Astronomers can combine the signals received from two or more radio telescopes and create an image that would be similar to one produced by a single telescope with a diameter equal to the distance between the two. In some cases, radio telescopes have been set up thousands of miles apart. Computers combine the information from the different telescopes to create a continent-wide observatory.

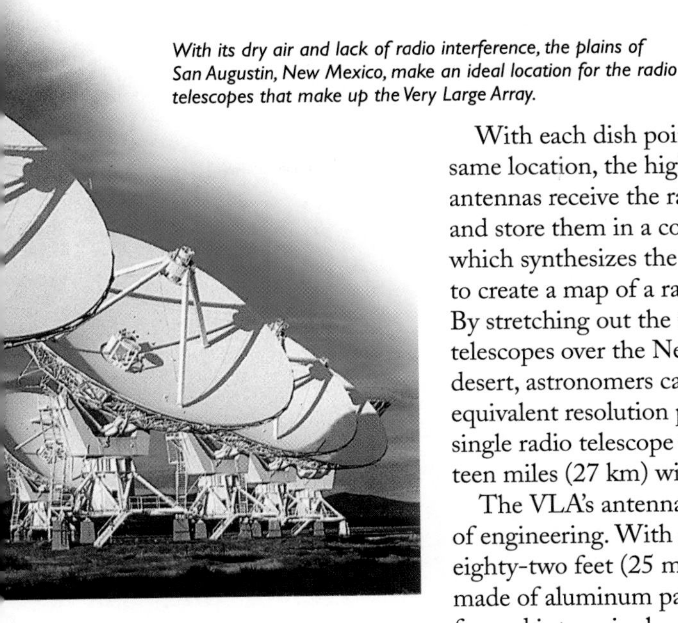

With its dry air and lack of radio interference, the plains of San Augustin, New Mexico, make an ideal location for the radio telescopes that make up the Very Large Array.

MANY VIEWS, ONE IMAGE

Because radio waves are one million times longer than those of visible light, radio telescopes tend to be much bigger than optical telescopes. A technique called interferometry permits them to function in groups, simulating an instrument that is larger than any of its parts. The VLA's 230-ton instruments, for example, are arranged in a Y pattern and can be moved around by special transporters for various observational purposes.

Depending on the level of detail required by researchers, the antennas can be clustered within a small two-thousand-foot- (610-m)-diameter area for a wide-angle view or relocated so that each arm is some thirteen miles (21 km) long. The bigger the distance between the antennas, the better the resolution.

With each dish pointing to the same location, the highly sensitive antennas receive the radio signals and store them in a computer, which synthesizes the information to create a map of a radio source. By stretching out the twenty-seven telescopes over the New Mexico desert, astronomers can create the equivalent resolution powers of a single radio telescope up to seventeen miles (27 km) wide.

The VLA's antennas are marvels of engineering. With a diameter of eighty-two feet (25 m), each dish is made of aluminum panels that are formed into a single, smooth, parabolic surface, which collects and reflects radio waves in the same way as the mirror of a reflecting telescope. The waves are focused to a detector, which transforms them into electrical signals that are then amplified to increase their strength.

Sensitive Receivers

By the time radio messages reached Earth from the *Voyager 2* spacecraft as it skimmed by Neptune, the signal had been reduced to one-billionth the power needed to run a digital watch. A group of radio dishes called the Deep Space Network was able to amplify the signals and process them into clear images.

Eyes on the Sky

Located in one of the most exotic spots on Earth, the twin Keck telescopes stare piercingly into the deepest recesses of space. The Kecks' perch, atop Hawaii's 13,800-foot (4,206-m) Mauna Kea, was chosen because of the smooth air flow over the mountain and the thin air that minimizes atmospheric distortion. Also, with no nearby large cities, the skies are clear of the dust and light pollution that can seriously obscure a telescope's view.

The two giant Keck telescopes, each standing eight stories tall and weighing three hundred tons, are the world's largest optical and infrared instruments. No telescope, not even the famed Hubble space telescope, can surpass their combined powers of resolution.

Much of the Kecks' success is due to their revolutionary design. Structures as massive as these are susceptible to the relentless pull of gravity, and even the slightest alteration in the frame could distort the sensitive main mirrors, rendering them myopic leviathans. To counteract this problem, engineers built each of the telescopes' thirty-three-foot (10-m) primary mirrors in thirty-six hexagonal sections. Each section is monitored by sensors that detect even the most minute distortion in their shape.

A complex system of precision pistons constantly adjusts each segment up and down and side to side, as often as twice a second. These corrections keep the mirrors aligned with an accuracy one thousand times thinner than a human hair and compensate for the telescope's continuous flexing as it tracks celestial objects across the sky. Thus, while each section of the Kecks' honey-combed surface moves independently, they all work in concert to create a single piece of near-perfect reflective glass.

NEW GALAXIES

Armed with these powerful telescopes, scientists have been able to probe the universe in unparalleled ways. The Kecks have been instrumental in the discovery of new

Location of the Kecks' telescopes atop the extinct volcano Mauna Kea places them higher than 35 percent of Earth's atmosphere.

Mirror, Mirror

Telescope mirrors are not always made of solid glasslike material. Astronomers have also been experimenting with mirrors fashioned from large shallow dishes of mercury. When such a dish is spun at a predetermined speed, centrifugal force molds the mercury into a reflective concave shape. While liquid mirrors cannot be tilted to track or aim at specific targets, they provide excellent optical quality for observing celestial objects overhead. This, combined with the fact that liquid mirrors cost roughly a tenth as much as equivalent solid-surface mirrors, is making these revolutionary devices viable observing tools.

galaxies and have proven invaluable in studying stars. In 1998 the Kecks gathered data on a distant supernova that suggested the universe is expanding at an ever-accelerating speed. This revolutionary finding contradicted the conventional theory that the expansion of the universe was gradually slowing down. The Kecks have also conducted a search for new planetary systems.

Currently, work is underway in northern Chile to create a telescope that will rival the Kecks. Slated for completion in 2000, the Very Large Telescope will consist of four telescopes which, like the twin Kecks, will work separately or in unison to create better resolution. The VLT's four components will each house a single 27-foot (8-m) mirror. When combined, they will have the light-gathering power of a 52-foot (16-m) mirror and produce ten times the resolution of the Hubble Space Telescope.

The Kecks' mirrors are made of a special glass-ceramic compound called Zerodur, which contains crystallized silica molecules. These crystals react to temperature changes in an opposite way to the glass component of the mirrors. The result is a counteracting effect that virtually eliminates expansion and contraction that would seriously compromise the performance of the telescope.

The Hubble

The saga of the Hubble Space Telescope (HST) has all the makings of a good Hollywood movie. Launched with great fanfare in 1990, Hubble initially was a crushing disappointment, but a series of space shuttle rescue-and-repair missions enabled it to fulfill its lofty promise.

Although the idea of an orbital telescope had been proposed as early as 1946, it wasn't until 1977 that work on the HST finally began. Like all telescopes, Hubble would only be as good as its eight-foot (2.4 m) primary mirror. Technicians took eight months to complete the painstaking task of polishing the surface, sometimes grinding away a mere millionth of an inch over the course of a week.

By the time they were finished. Hubble's mirror was so smooth that if it were enlarged to the size of the Gulf of Mexico the largest surface irregularity would be a ripple less than one fifth of an inch (0.5 cm) in height. However, a disastrous mistake had been made.

Soon after HST was placed in orbit, scientists realized that the mirror suffered from "spherical aberration." Due to a technical glitch, the mirror had been ground to the wrong shape. It was, as one engineer described it, "perfectly wrong."

OUTER SPACE REPAIR CALL

The problem was rectified in December 1993, when astronauts aboard the space shuttle *Endeavour* fitted Hubble with three coin-sized corrective mirrors. The adjustment worked better than expected and the

Silicon Magic

Like images taken by all modern telescopes, those from Hubble are not recorded on film. Instead scientists rely on a charge-coupled device (CCD), a wafer-thin silicon chip that transforms light falling onto it into electrical charges. A CCD is divided into hundreds of thousands of tiny picture elements, or pixels. When a photon of light strikes a pixel, it is converted into electrons, which move from one pixel to the next in conveyor-belt style until they reach an output register. There they are counted and converted into an image by computer. Using CCD technology, a half-page color photograph might be made up of more than two million pixels.

The Hubble Space Telescope (left) *has undergone several maintenance and equipment upgrades. Extensive work, such as the 1993 mission to repair Hubble's primary mirror, is done by bringing the telescope inside the shuttle's cargo bay. Less complicated tasks, such as repairing frayed insulation, can be completed on space walks.*

The improvement provided by the lenses that repaired Hubble's mirror problem is evident in this comparison of an old image of M100 (left) *and the newer, sharper one of it* (above)*, taken after the repair mission. Hubble's images are some ten times sharper than those produced by conventional, ground-based equipment.*

first images transmitted by the orbiting observatory thrilled astronomers with their clarity.

Looking through Hubble's eye, scientists have watched as young stars develop and old stars die. In November 1998 the bus-sized observatory recorded images of thousands of previously undiscovered galaxies. Situated twelve billion light-years away on the very fringe of the observable universe, some of these ancient galaxies have given scientists glimpses of what the universe looked like soon after the Big Bang.

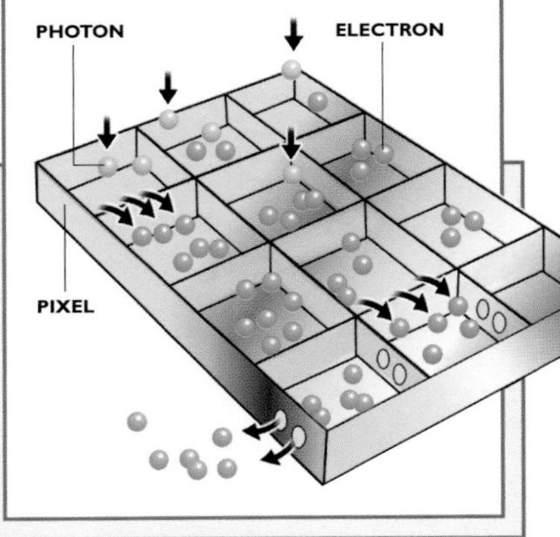

One of Hubble's great strengths is its versatility. Already one of the most powerful visible-light telescopes in the world, it also does work in the ultraviolet and infrared spectra. Space shuttles periodically link up with Hubble to upgrade the existing equipment with cutting-edge instruments such as new cameras and imaging spectrographs. Following its last maintenance rendezvous in 2002, Hubble will continue gathering vital data until its operational life comes to an end in 2010.

High-Energy Observation

X-rays and gamma rays are high-energy invisible forms of radiation at the upper end of the electromagnetic spectrum. Both sorts of rays are produced by violent cosmic events, such as exploding stars and cosmic matter swirling around black holes. Once undetectable, these high-frequency waves are now being mapped and analyzed by scientists armed with a new generation of space telescopes.

HIGH-TECH OBSERVATIONS

Scheduled for launch in the summer of 1999, the Chandra X-Ray Observatory is an orbiting telescope capable of capturing images fifty times more detailed than those of its predecessors. At the heart of Chandra lie four nesting mirrors that were made by hollowing out solid cylinders of glass. Polished to a fine sheen, the glass was then treated with iridium, a highly reflective rare metal. The mirrors are angled slightly so that the X-rays ricochet between them and then converge at a focal point where the detectors of the telescope are located.

X-rays come in varying wavelengths depending on the conditions in which they were produced. When a star explodes into a supernova, for example, it creates an inferno as bright as a billion stars and fires off super-heated gas that emits X-rays. Chandra can examine these high-energy emissions to analyze what gases and heavy elements were present at the time of the explosion. By studying similar X-rays from young stars, scientists hope to gain insight into the birth and development of our sun and how the elements necessary for life were created. Chandra will also search for clouds of intergalactic dust—one possible form of dark matter in the universe *(page 19)*.

DETECTING GAMMA RAYS

What Chandra is to X-ray research, the Compton Gamma-Ray Observatory is to gamma rays.

Sophisticated X-Ray Telescope

The tube-shaped mirrors of the Chandra X-ray telescope bounce incoming X-rays to detectors thirty-three feet (10 m) away. The mirrors are layered to boost Chandra's sensitivity. The telescope was named after American astrophysicist S. Chandrasekhar.

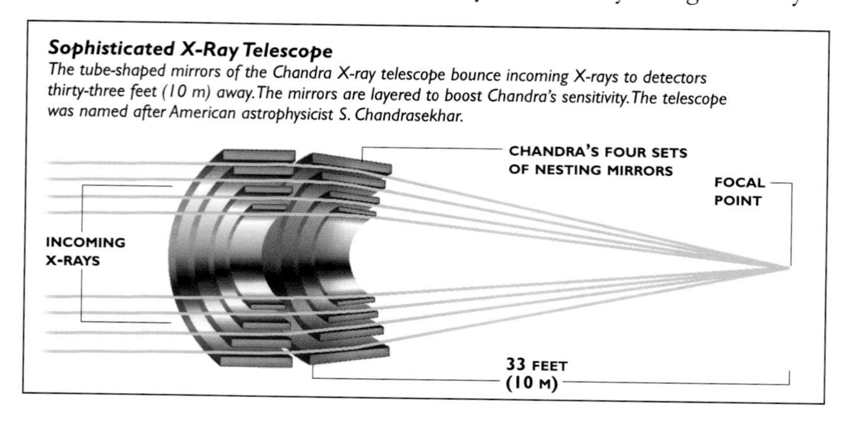

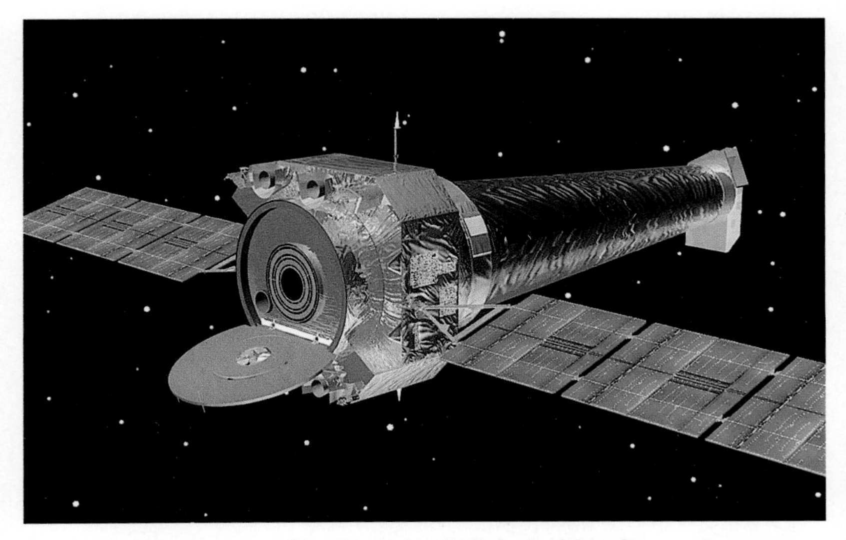

The Chandra X-Ray Observatory (above) *is third in NASA's family of Great Observatories program, which also includes the Hubble Space Telescope and the Compton Gamma-Ray Observatory* (below).

Scientists believe that these rays, products of the most explosive forces in the cosmos, will shed light on the cataclysmic events that formed the universe.

In low-orbit around Earth since 1991, the seventeen-ton observatory has been gathering data twenty-four hours a day. Included among its high-tech tools is the Burst and Transient Source Experiment (BATSE). This telescope scans space in search of short, often elusive, bursts of gamma rays.

In January 1999 Compton helped scientists make the first eyewitness record of a distant gamma-ray burst. Recorded by BATSE, the pulse of radiation produced the energy equivalent of ten million billion suns in a matter of seconds.

Compton is also providing new insights into pulsars—rotating stars as small as six miles (10 km) in diameter, and the extremely bright remote galaxies known as quasars. In addition, the observatory has been instrumental in the discovery of new sources of gamma rays being emitted from the central core of distant galaxies—the product of black holes as they swallow surrounding dust and gas.

Beyond the Electromagnetic Spectrum

The two 2.5-mile- (4-km)-long arms of the Laser Interferometer Gravitational Wave Observatory (LIGO) stretch out in Livingstone, Louisiana. Two sites more than two thousand miles (3,219 km) apart were chosen for LIGO to eliminate discrepancies caused by local disturbances such as minor earthquakes.

Most of what scientists know about the universe they have gleaned from the study of electromagnetic waves—the spectrum extending from radio waves to gamma rays. But since only about 10 percent of all matter can be observed this way, astronomers are continually searching for new windows on the cosmos. Paradoxically, some of those windows are underground.

EINSTEIN'S LEGACY

Although Albert Einstein first proposed the existence of gravitational waves, ripples in the fabric of spacetime, as far back as 1916 *(pages 14 to 15)*, detecting these forces has proven to be an impossible task. The Laser Interferometer Gravitational Wave Observatory (LIGO) may soon change that.

Slated to open in the year 2000, LIGO will consist of two observatories: one in Louisiana and another in Washington state. Each of LIGO's sites will consist of a V-shaped vacuum pipe, each arm of which is 2.5 miles (4 km) long. A series of mirrors will be hung from wires at the end of each arm and at the vertex.

Scientists intend to bounce a laser beam back and forth between the mirrors in the arms of the observatories. LIGO's detectors are so sensitive they will be able to measure any change in the distance between the mirrors—changes that could be as minor as one-hundred-millionth the diameter of a hydrogen atom. The detectors will then record and measure any anomaly or disturbance, perhaps detecting gravity waves for the first time.

THE SEARCH FOR NEUTRINOS

Another subterranean marvel is Canada's Sudbury Neutrino Observatory (SNO) near Sudbury, Ontario. Situated in a nickel mine some sixty-eight hundred feet (2,100 m) below the Earth's surface, SNO is designed to detect one of the most elusive particles in the universe: neutrinos.

Neutrinos are subatomic particles produced in huge quantities by the nuclear activity of stars. Scientists believe they are so numerous that they may constitute the majority of the universe's mass and thus account for much of the gravitational glue that helps bind the universe together. Until now, scientists were unsure if neutrinos had any mass at all. However, recent experiments at the Super Kamiokande Dectector in Japan have indicated that they do.

Neutrinos are difficult to detect directly because they rarely interact with matter and travel close to the speed of light. (A neutrino could slip through a piece of lead one light-year thick without even slowing down.) Scientists therefore try to observe them indirectly through their occasional interaction with the subatomic particles in a clear liquid, where the event can be more easily detected and measured.

SNO was built for this purpose. At its core sits a giant spherical tank that holds one thousand tons of "heavy water," molecules of oxygen and a hydrogen isotope that contains a neutron in its nucleus in addition to a proton.

As neutrinos bombard the Earth, they slip through the ground and into the tank. When neutrinos collide with the heavy-hydrogen atoms, they produce flashes of blue light that are registered by the ten thousand light detectors surrounding the tank. Although trillions of neutrinos will pass through the observatory each day, scientists expect that only twenty or so will be detected.

Scientists hope that SNO will give them valuable insight into the properties of neutrinos, and the role they play in determining the evolution of the universe.

The photograph at left shows the bottom of the Sudbury Neutrino Observatory's forty-foot- (12-m)-diameter acrylic vessel, surrounded by a partially completed photo sensor sphere. The sensors will attempt to detect the interaction of neutrinos with the thousand tons of heavy water contained by the two-inch- (5-cm)-thick walls of the tank.

Making Contact

In 4 B.C., the Greek philosopher Metrodorus wrote, "To consider the Earth as the only populated world in infinite space is as absurd as to assert that in an entire field of millet, only one grain will grow." Two thousand years later, the words still ring true to a growing group of people who gaze skyward, hoping to contact other worlds.

The SETI (Search for Extraterrestrial Intelligence) movement began in 1960, when a young astronomer named Frank Drake began to scan the sky with a radio telescope dish looking for signals being transmitted from nearby stars. Although Drake failed to uncover alien life, his methods established the strategy for the first decade of SETI research, namely to aim a large radio telescope at a cluster of stars and wait for a signal.

In 1974 Drake and colleague Carl Sagan developed a more aggressive approach. Stationed at the Arecibo Observatory in Puerto Rico, the American astronomers used the installation's one-thousand-foot- (305-m)-diameter radio telescope dish to send

The SETI@home project joins the forces of professional and amateur astronomers. Data recorded by the Arecibo telescope (right) *is parceled up and sent to SETI@home members via the internet. Using a specialized analytical software in the form of a screensaver* (below), *participants can process a huge volume of data that would otherwise take years to analyze.*

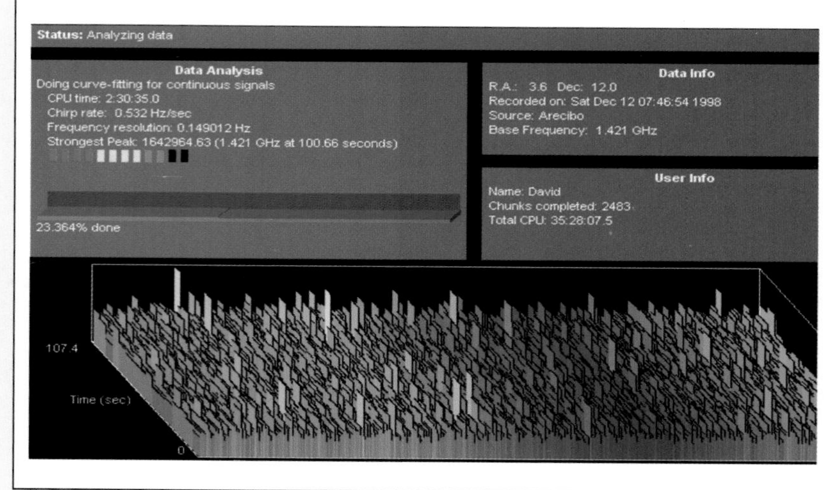

a message to M13, a globular star cluster twenty-five thousand light-years away. Written in binary code, the message could be decoded into illustrations of the double helix of DNA—the fundamental key to human heredity—and of a human and the Arecibo dish itself.

Scientists also took another tack. Rather than beaming out messages to distant planets, they decided to attempt to transport something tangible to these worlds. In 1977 NASA launched *Voyager 1* and *Voyager 2* on a tour of the planets and beyond. Both craft are equipped with a special phonographic record, a needle, and coded instructions on how to operate it. Extraterrestrial beings who play the record will be welcomed in fifty-five languages and serenaded by a myriad of Earth sounds, including a howling dog, a crying baby, a Mozart aria, a kiss, and crashing surf.

It will take the two Voyager *probes forty thousand years to pass within a light-year of another star. The recorded greetings attached to the crafts have been fitted with a protective gold-plated aluminum cover.*

SETI TODAY

More than twenty years after transmitting that first interstellar message, the Arecibo Observatory is still at the center of SETI operations. SERENDIP III is capable of monitoring 4.2 million frequencies every 1.7 seconds with a detector mounted on the telescope. The next generation of instruments, being developed for SERENDIP IV, promises an even greater reach by being able to analyze forty times as many frequencies as its predecessor.

In the past, governmental budget cuts have sounded the death knell for numerous SETI programs. Ironically, though, these setbacks may have actually strengthened SETI by galvanizing a worldwide corps of enthusiastic amateurs. Armed with personal computers and affordable satellite dishes, home-based amateur astronomers are picking up where the professionals have had to leave off. Now, instead of hoping for success from a single radio telescope analyzing a single star or star cluster, organizations such as the SETI League are scanning space with hundreds of smaller personal devices.

Should members of SETI receive a signal from an extraterrestrial, they must follow strict verification programs and protocols in sharing their findings with scientists and the United Nations. To date, these amateur astronomers have received some heart-stopping, albeit false, alarm signals. In 1996 a pair of British amateurs picked up a strange signal they couldn't identify. Two weeks later they learned that their "extraterrestrial" was actually a classified U.S. Navy satellite.

SECTION TWO

The Backyard Astronomer

CHOOSING YOUR EQUIPMENT

When newcomers to astronomy think about equipment, they often have telescopes in mind. But there are other devices and accessories that can greatly add to the enjoyment of looking at the stars.

Skywatching offers endless pleasure for the amateur astronomer, but proper equipment is essential.

Because they are affordable and portable, binoculars *(pages 68 to 69)* are the perfect instruments for becoming familiar with the night-time sky. Not only do they magnify objects, but their wide field of view makes them useful for scanning large sections of the heavens.

The next step for most amateurs usually involves purchasing a telescope *(pages 70 to 73)*. This is a major investment, and it's well worth taking the time to make the right choice.

Although binoculars and telescopes represent the two most expensive big-ticket items for the beginner and keen amateur, they represent only part of the equipment picture. For starters, you need to know how to find objects in the night sky. A planisphere is one solution. Basically a miniaturized map

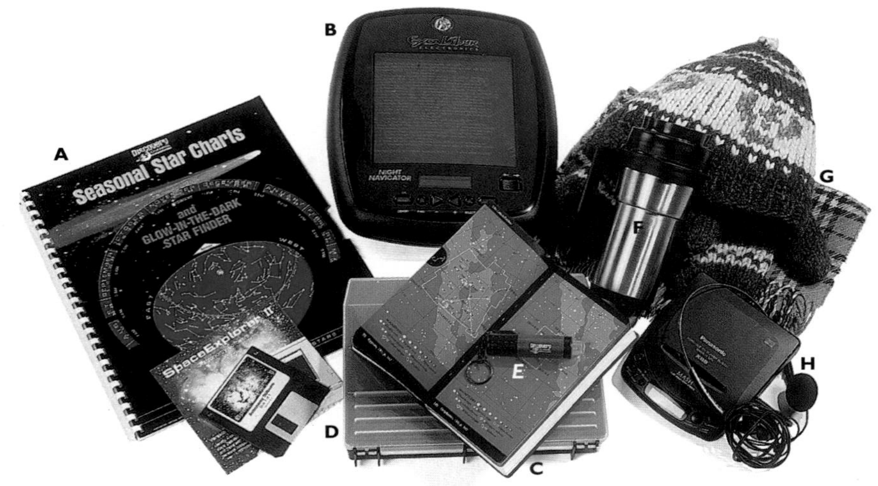

The right gear can improve your night-time observing sessions: (from left) *(A) a planisphere and (B) an astro-computer program for pinpointing celestial objects; (C) a handbook to the heavens; (D) a tool kit with basic tools such as wrenches and screwdrivers for adjusting a telescope; (E) a red flashlight that won't affect your dark-adapted eyes; (F) a thermos; (G) a hat and gloves; and (H) a portable cassette or CD player.*

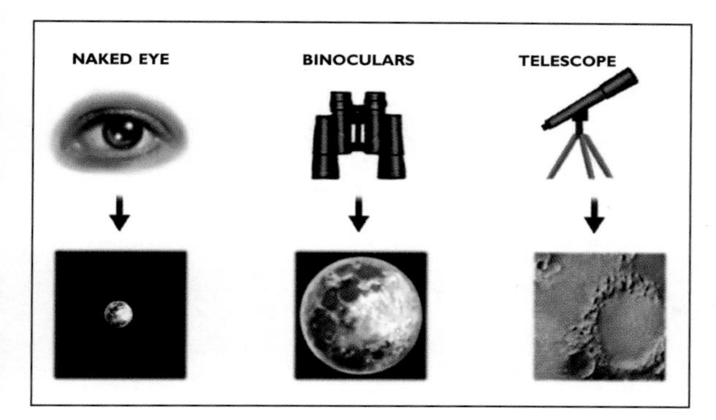

Fields of view
The unaided eye provides a 64-degree view of the sky; a pair of binoculars, 7 degrees, and a telescope, less than 2 degrees. Each has its advantages. The naked eye offers a wide view of the sky—ideal for spotting meteors and observing constellations. Binoculars provide some magnification, telescopes, much more still. But that power comes with a cost—an increasingly limited field of view.

of the sky that rotates on a small disk, planispheres are available for different latitudes and provide a month-by-month view of the sky for any date and time.

A good star atlas offers more detail. These books are characterized by the faintest-magnitude stars they show. A sixth-magnitude atlas shows all objects down to magnitude 6—approximately the limit of the unaided human eye (eighty-five hundred stars). The ninth-magnitude atlas *Uranometria 2000.0* plots more than three hundred thousand stars. The fainter the stars shown, the smaller the area of sky that appears on each map.

Computers are an increasingly important tool in amateur astronomy. Certain astro-computer programs can link telescopes to PCs so when the user specifies what celestial object he wishes to examine, the computer moves the telescope to the exact coordinates. The same programs will supply users with on-screen information regarding the bodies they are observing. Star maps and charts of the solar system have also been added to the computer's ever-growing astronomical repertoire.

> *"No one regards what is before his feet; we all gaze at the stars."*
>
> — Quintus Ennius
> 239-169 B.C.

DRESS FOR SUCCESS

For optimum comfort in the field, many people like to use observing chairs and tables. The best chairs are actually lightweight adjustable stools that fold for easy storage. Folding tables are perfect to hold charts, logbooks, and star atlases.

Although it seems obvious, some of the most important—and most neglected—equipment is proper clothing. Astronomy is an activity in which participants spend long periods of time relatively motionless. It is easy to get chilled, especially on autumn or winter nights. The general rule of thumb is that dressing in a series of thin layers traps more warm air than a few thick ones. Also, you are only as warm as your coldest part. Wear hats, gloves, and long underwear when the temperature merits them.

Selecting and Using Binoculars

Of all the equipment available to the amateur astronomer, binoculars are the most versatile and most essential. They provide an invaluable bridge between unaided eyes and telescopes in terms of power and field of view—the width of sky the binoculars show.

The main considerations in selecting binoculars for astronomy are optical quality, weight, price, magnification, and objective-lens size. While any type and size of binoculars will do the basic job, the optics in higher-quality models best show faint stars and deep-sky objects.

Understanding general concepts and the special terms related to binoculars is critical to making the right purchase. Binoculars have two numbers engraved near the eyepiece—for example, 11x80. The first number is the power, indicating the magnification of the view. The second number is the aperture, the diameter of the front, or objective, lenses in millimeters. So, 11x80 means eleven power and eighty-millimeter objective lenses. The larger the objective lenses, the brighter the object will appear.

On a moonless night, 7x50 binoculars can detect more than 150,000 stars, fifty times the number visible to the naked eye. With the relatively modest investment of a pair of binoculars, the night sky reveals a wealth of treasures, including craters and mountains on the Moon, nebulae, more than a dozen galaxies, star clusters, the planets Uranus and Neptune, numerous double stars—in fact, years' worth of interesting celestial sights.

There are two basic types of binoculars: porro prism and roof prism. Porro-prism binoculars feature four reflections—that is, interi-

Porro-prism binoculars

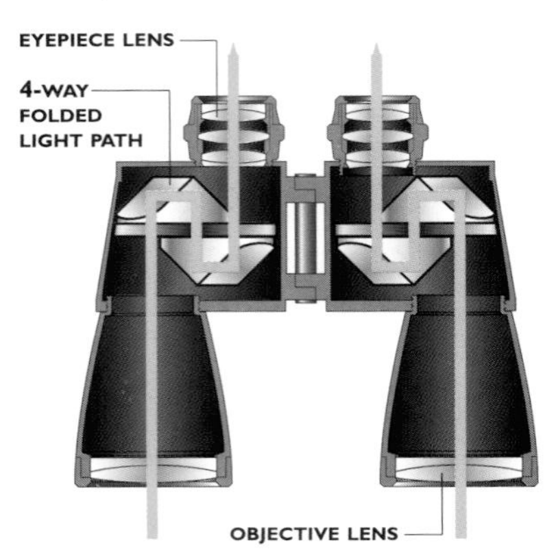

Roof-prism binoculars

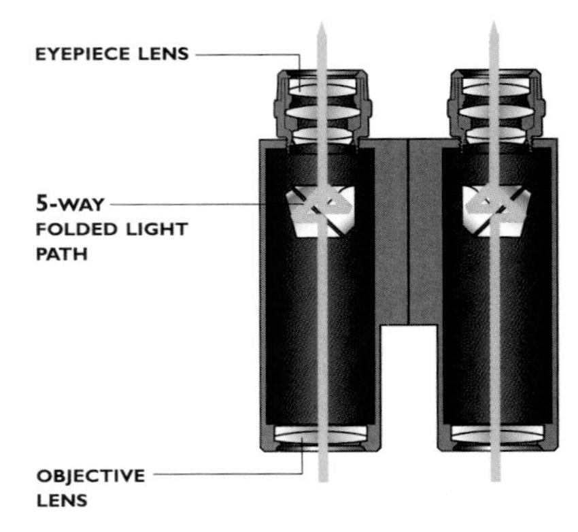

Two Types of Binoculars
The two basic types of binoculars used by astronomers are porro prism (left) *and roof prism* (right).

While the magnification they provide can't compete with a telescope, binoculars offer a much wider field of view and a more relaxing view of the night sky, since the brain finds it easier to process information from two eyes than just one. In general, 10-power binoculars are about as powerful as most people can hold.

or prisms fold incoming light four times. This, combined with their large objective lenses, means that lots of light reaches the eye, making these binoculars a good choice for low-light conditions. The large lenses of the porro-prism binoculars also provide a wider field of view.

Roof-prism binoculars reflect light five times, bringing less light to the eye. These binoculars are small and light, but tend to be more expensive to manufacture.

Porro-prism binoculars have long been the standard and, when all factors are considered, are an excellent choice for astronomers.

Binocular users who wear glasses for astigmatism should consider buying long eye-relief binoculars. This model is designed to push the exit pupil—the optimum position for the eye—to twenty-eight millimeters from the surface of the exit lens, compared with ten to fifteen millimeters for normal binoculars, making it much easier to use glasses. If you're only modestly near- or far-sighted, however, just take off your glasses and focus the binoculars to compensate.

A Steadying Influence

Binoculars present a rich wide-field view of the night sky, but holding a pair steady for lengthy viewing can prove taxing on neck and arm muscles. The problem is compounded with greater magnification: The higher the power, the greater the weight and the more difficult it is to hold binoculars steady. One solution is to invest in a good-quality binocular mount and tripod. The model shown here, with its counterweight, quick-release leg locks and simple adjustable height mechanism, can free up your hands and provide a steady view for as long as you like.

Choosing the Right Telescope

Telescopes have come a long way since Galileo explored Jupiter and its satellites with his 30-power scope in the early 1600s. Galileo's device, a tiny, hand-held spy glass, bears little resemblance to the powerful high-tech instruments on the market today. In fact, the sheer variety of telescopes can make it difficult to choose the most appropriate model.

Despite what many neophytes may think, bigger isn't necessarily better. It is fairly common to find large, powerful telescopes languishing in someone's basement because they were just too big and awkward to carry or use. This is why a person's first telescope should combine such qualities as clear optics and a steady mount with such practical considerations as ease of use and portability.

THREE MAIN DESIGNS

There are three basic types of telescopes available for the amateur: refractors, reflectors, and Schmidt-Cassegrains. Each type has its strengths and weaknesses.

Refractors use lenses for their optics. When lightwaves pass through the instrument's large glass lens, also known as the objective lens, they are bent and focused to a point down the telescope's body where they can then be viewed through the eyepiece. This image can be magnified to varying degrees, depending on the strength of the eyepiece.

The advantage to a refractor is that its tube is closed so it generally is less affected by distorting air currents than a reflector, which is open at one end. A refractor also does not suffer from the loss of light, called diffraction, that is caused by a reflector's secondary mirror and its supports. On the negative side, well-made refractors are expensive. Inch-for-inch, you can expect to pay considerably more for a refractor than for a reflector of similar quality.

The first reflector was built by Sir Isaac Newton in 1668. Instead of using a glass lens to collect and focus incoming lightwaves, Newton outfitted his telescope with mirrors. Light

NEWTONIAN REFLECTOR WITH EQUATORIAL MOUNT

DOBSONIAN REFLECTOR AND MOUNT

entering the front end of a Newtonian reflector is collected by the concave primary mirror located at the back of the scope's body. The light is then bounced back up the body to a smaller flat mirror. This secondary mirror is angled so that it reflects the light rays to the eyepiece.

While the reflector's mirrors do away with some optical problems, they do have drawbacks. Because light suffers interference as it passes by the central secondary mirror, the image produced by a a reflecting telescope is not quite as crisp as that of a refractor. But because the telescope is less expensive than a refractor, an amateur astroomer can simply buy a little more aperature for extra light-gathering power.

The third type of telescope, the Schmidt-Cassegrain, is somewhat of a hybrid because it uses a combination of a main and a secondary mirror, along with a corrector plate that sharpens the image. These telescopes tend to be more expensive than similar-sized Newtonians, but they are also much more portable, and have become the mainstay for serious backyard astronomers.

SOME MORE CONSIDERATIONS

No matter what type of telescope a person is buying, two factors should be kept in mind: aperture and focal ratio. The larger a telescope's aperture—the diameter of its objective lens or mirror—the more light it collects, which means that scopes with

Telescopes and Their Mounts

Telescopes come in different models: refractors, reflectors, and Schmidt-Cassegrains; with two mounting methods: equatorial and altazimuth. The equatorial mount can track celestial objects with a single movement, but has to be aligned first with the north celestial pole. Altazimuth mounts need to be moved both horizontally and vertically, a more challenging task, although the Dobsonian version tends to produce lower power so it is relatively easy to nudge the scope along. Computer-assisted altazimuth mounts automatically track an object, but they are expensive.

REFRACTOR WITH ALTAZIMUTH MOUNT

SCHMIDT-CASSEGRAIN REFLECTOR WITH COMPUTER-ASSISTED ALTAZIMUTH FORK MOUNT

big objective lenses or primary mirrors are better for viewing dim objects. The focal ratio is the ratio of a telescope's aperture to its focal length—the distance that light must travel between the primary mirror or objective lens to the eyepiece. For example, a telescope with a focal length of 1,200 millimeters and a 150-millimeter mirror would have a focal ratio of 1,200/150, or f/8. Generally, a telescope with an f/ratio between f/6 and f/8 will prove most versatile for backyard astronomers.

Different observing calls for different telescopes. If you're searching for faint objects, buy as much aperture as you can afford. This probably means a reflector. But make sure it is not so large that you cannot transport it out to the dark skies where it will perform best. On the other hand, if you plan to spend your time looking at the moon and the planets, consider buying a high-quality refractor with at least a 75 millimeter aperture. In any case, avoid buying a telescope that is advertised by its magnification alone—the ones you often seen in department stores. They tend to be of poor quality.

CHOOSING EYEPIECES

Essentially just short focal-length lens systems, eyepieces provide different magnifications and different fields of view. The higher the magnification, the smaller the field of view. Try to look for eyepieces that have been multi-coated on all surfaces, and replace any that come with your telescope that aren't. To figure out the magnification that an eyepiece provides, divide the eyepiece's focal length into that of the telescope's. For example, a 1,500-millimeter focal-length telescope with a 25-millimeter eyepiece will provide 1,500/25, or 60 power.

BUILD A SOLID BASE

Surprisingly, while many backyard astronomers are willing to make a significant investment in a top-grade telescope, they often sabotage their efforts by buying a low-quality mount and tripod. The best mounts are not only stable, they also move smoothly as they track objects across the sky. Many experts recommend altazimuth mountings for beginners because they are easy to set up and use. These mounts have controls that move the scope up and down and side to side. The term altazimuth refers to the vertical and horizontal

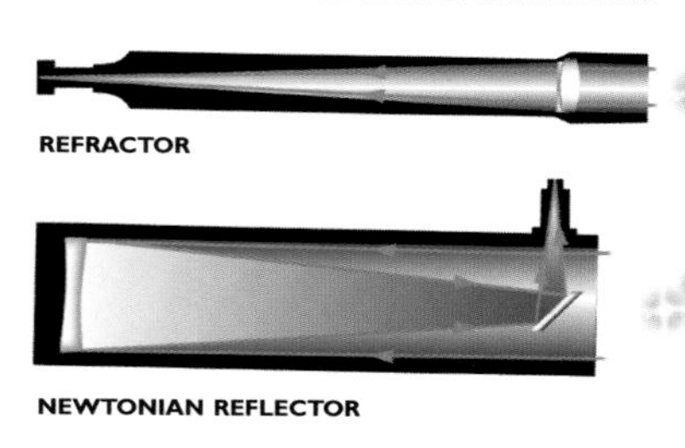

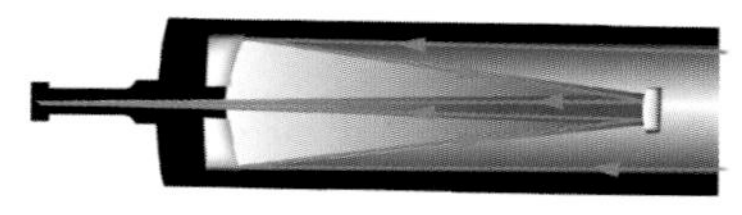

SCHMIDT-CASSEGRAIN REFLECTOR

Three Telescopes, Three Light Paths
Light entering a refractor telescope (top) *is bent inward by the objective lens and brought into focus. A reflector* (center) *bounces light from the primary mirror to the angled secondary mirror and then straight up to the eyepiece. With the Schmidt-Cassegrain reflector* (bottom), *light passes through a corrector plate, then reflects from the primary mirror off the secondary mirror and back through a hole in the primary to the eyepiece.*

KELLNER
f/4.5 and higher

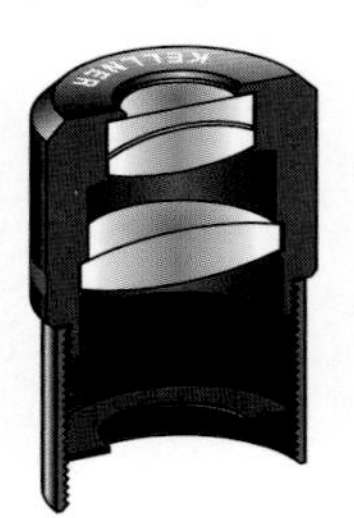

ORTHOSCOPIC
f/4.5 and higher

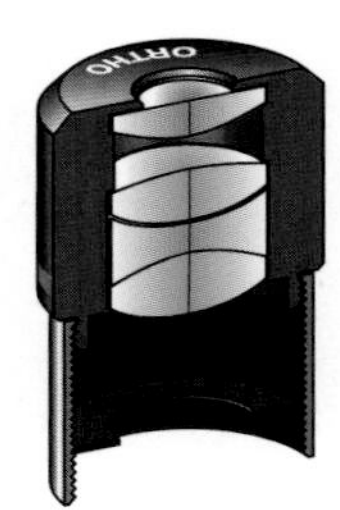

PLOSSL
f/4.5 and higher

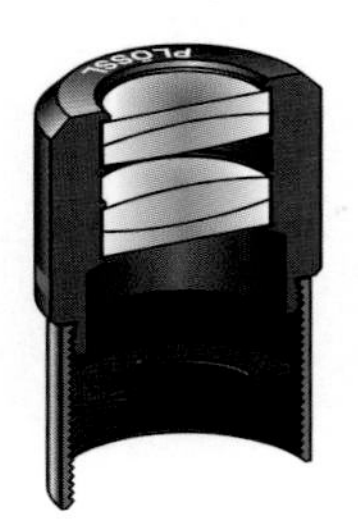

ERFLE
f/6 and higher

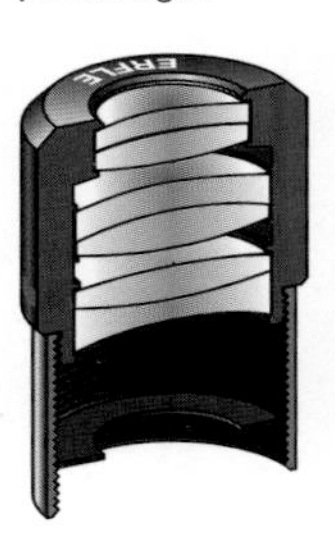

Essential Eyepieces
In general, eyepieces have two or more lenses. The lens farthest from the eye is called the field lens, while the one nearest the eye is called, logically, the eye lens. Eyepieces come in a variety of models. Above are some of the more popular eyepieces on the market today, complete with the f/ratios that they work best with. Make sure your telescope comes with an eyepiece barrel 1.25 inches (3.175 cm) or larger in diameter; 0.965-inch (2.45-cm) barrels are a hallmark of lower-quality instruments.

reference points, altitude and azimuth, that can be used to pinpoint a celestial body's location in the sky *(page 88)*. Dobsonian mounts, a type of altazimuth, are a good choice: They are very stable and smooth in operation.

The second option is the equatorial mounting. Rather than moving straight up and down or side to side, these mountings are aligned with the Earth's axis so the telescope moves parallel to the lines of declination and right ascension *(page 89)*. As such, they track stars as they move across the sky with one motion.

Whatever mount you choose, you need a good finderscope. Mounted on the main scope, this small low-powered scope is ideal for scanning large sections of the sky in search of particular objects. Get at least a 6x30 finder; better yet, an 8x50.

COMPUTERS TO THE RESCUE

Computerized telescopes have alleviated much of the complexity of finding and tracking objects in the night sky. Using two bright stars as its reference points, a computer-driven telescope tracks the sky by automatically making adjustments for the Earth's rotation. In addition, the computers have extensive databases of objects in the night sky. By simply pushing a couple of buttons on a hand-control unit, you can locate and lock onto the star or galaxy you want to see.

Never look at the sun through a telescope without taking precautions, such as using a solar filter, a device that is fitted to the objective lens or primary mirror. See page 75 for more information on viewing the sun safely.

FIRST LIGHT

Telescopes can transport an observer to neighboring planets or distant galaxies. But first, they have to be set up properly, and that involves a little more than simply aiming them at the night sky.

You've unpacked your new telescope and placed it lovingly under the stars. It's a moment charged with anticipation. Will it show you wonders? Or will you simply wonder what's gone wrong?

DAYTIME CHECKOUT

To avoid disappointment, never use your telescope for the first time at night. Instead, assemble it indoors. Learn how the telescope moves and what each of its controls does before heading outdoors.

If you have an equatorial mount, set the angle of the polar axis, the axis around which the telescope (and counterweight, if it has one) swings from east to west. German equatorial mounts (used on refractors and reflectors) and the wedges that tilt Schmidt-Cassegrains all have provisions for tilting the polar axis to an angle equal to the latitude of your observing site. Many have protractor-like scales to help find the correct setting.

Two other key adjustments are best done outside during the day. First, insert the lowest-power eyepiece (usually marked 40 mm or 25 mm). Now sight along the tube to aim at a target that is at least a mile (1.6 km) away. Rack the focuser in and out until you get a sharp image. The telescope is now ready for use at night.

Second, adjust the screws holding the small finderscope so it is centered on the same target as the main telescope. With its finderscope set, your telescope is ready for its baptismal use.

FIRST NIGHT

Telescopes must be used outdoors; trying to look through a window (either open or closed) from inside a warm room

Aligning the Polar Axis
An equatorial mount is correctly polar-aligned when its polar axis is parallel to the Earth's axis of rotation. Accurate alignment requires the polar axis to point directly to the north or south celestial pole, depending on the observer's hemisphere. However, aligning on Polaris in the north (right) *and Sigma Octantis in the south will do for most general observing.*

"O, telescope, instrument much knowledge more precious than any sceptre, is not he who holds thee in his hand made king and lord of the works of God?"

— Johannes Kepler (1571-1630)

yields fuzzy images. Allow fifteen to thirty minutes for the telescope to "settle down." Only when it has cooled to the night air will it provide the sharpest images.

Equatorial mounts need to be oriented so that the polar axis aims due north, toward Polaris, the North Star. This may require no more than tilting up the whole telescope and turning it. Then, adjust the leg heights to level the telescope as best you can. If you've set the polar-axis angle correctly, the scope should now be aligned well enough for casual stargazing.

For your first views, always use the lowest-power eyepiece, with no power-doubling Barlow lenses. High magnification makes it hard to find targets, and often produces blurred images.

Safe Daytime Stargazing

Viewed with the proper techniques, the Sun can provide an ever-changing disk, pock-marked by dark Sunspots.

The preferred method is to use a solar filter that slips over the front of the telescope tube. Made of Mylar or metal-coated glass, these filters reduce the light to safe levels before it enters the telescope. Because they can crack, "Sun filters" that screw into eyepieces are too dangerous and shouldn't be used.

Another technique is to project the Sun's image onto a white card held behind the eyepiece. Because this requires aiming an unfiltered telescope directly at the Sun, it can be risky for inquisitive children and telescope optics.

In either case, make sure you remove or block the finder scope to prevent accidental exposure to its concentrated focus.

You can also make a solar viewer from a small box *(page 178)*.

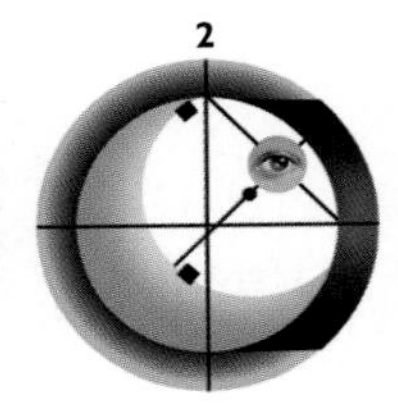

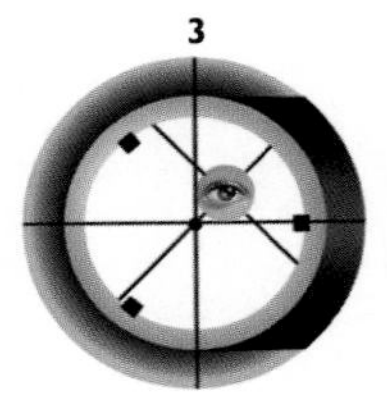

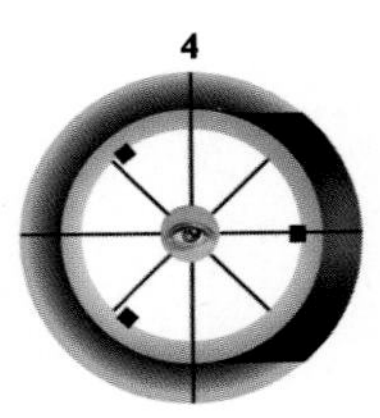

Collimating Your Optics

To yield the sharpest views, the secondary diagonal mirror and the main mirror in a Newtonian telescope sometimes need adjusting, or collimating, to make them perfectly parallel. Here's how to do it.

1. If your telescope is badly collimated, this is what you'll likely see. 2. Now, adjust the vanes and central bolt holding the small secondary mirror to center this mirror under the focuser. 3. By turning its three screws, tilt the secondary to center the reflection of the large primary mirror. 4. Adjust the primary's three screws to center the reflection of the secondary mirror in the reflection of the primary.

On Schmidt-Cassegrain telescopes only the secondary mirror is adjustable. Sight a star at high power, slightly out of focus. Make small tweaks of the secondary's screws until the donut-shaped star image becomes as round and symmetrical as possible.

An Evening Out

While stargazing can be done almost anywhere—even from an urban backyard or a high-rise balcony—the best views of deep-sky objects such as galaxies, nebulae, and star clusters are obtained as far from city lights as possible.

Your best bet for finding a country site may be the local astronomy club. They likely have a site at a public park or private observatory. Viewing with a group also provides security and a lot more fun.

The ideal site will combine a host of desirable qualities. Below are some of the factors to keep in mind.

- Find a location where there is no major city and no sky glow to the south.
- It's best if no lights are visible, or are far away on the horizon.
- A buffer of trees is useful for hiding lights and blocking wind.
- Choose a spot that is not likely to be visited by nighttime revelers, but make sure that help is not too far away should you need it (if your car breaks down, for example).
- Try to find a spot that is easy to get to—and get out of—especially in winter.
- Stick to high and dry ground. Bug-infested swamps or foggy valleys are a disaster.
- Pick a place that is not likely to arouse suspicions of neighbors or officials.

Before you head into the field, take a little time to get organized. Collect all eyepieces and accessories into one durable camera case. Include any tools required to tighten loose screws. And if you need to plug into a car battery, don't forget power extension cords.

URBAN ASTRONOMY

Just because you can't afford the time to drive to that perfect country viewing site doesn't spell the end of your nighttime observing

Too Much Light

When the lights of a city spill into the sky and obscure the faint glow of celestial objects, light pollution results. And since light pollution means wasted energy, it's not just a problem for astronomers; it's a costly problem for everyone. In North America alone, light pollution results in two billion dollars of wasted energy each year.

In 1987 Arizona astronomers David Crawford and Tim Hunter founded the International Dark Sky Association in an attempt to stem the swelling tide of light pollution. It's a daunting, seemingly overwhelming challenge, but so far there have been some successes.

In 1994 the city of Tucson in Arizona instituted a strict policy on shielding outdoor lighting within thirty-five miles of nearby Kitt Peak National Observatory and banned the illumination of outdoor advertising between the hours of 11 p.m. and sunrise. The city has also switched to sodium lighting, which is both more electrically efficient and emits a diffuse yellow light that can easily be filtered out by astronomers.

A good observation spot, with as many of the qualities mentioned on page 76 as possible, can add greatly to the number of celestial objects visible to the amateur astronomer.

sessions. Telescope views of the moon and planets are unaffected by city lights. The prime requirement for a city site is to be shielded from nearby streetlights glaring into your eyes. Also, avoid looking directly over chimneys and heating vents. Rising heat waves distort planetary images. Nebula, or light pollution, filters block out yellow and blue light, while letting though the green and red wavelengths emitted by gaseous nebulas.

No matter where your observing site is, remember to take at least twenty minutes to allow your eyes to dilate as much as possible. An extra trick is to look a little to one side of the target. This "averted vision" places a faint nebula or galaxy on the more light-sensitive area of the eye's retina where it will pop into view.

The Art of Telescope Maintenance

Cold weather rarely harms the optical or mechanical parts of a telescope. But before bringing a cold telescope inside, cap it to prevent warm air from condensing on the optics and leaving a filmy residue.

To clean eyepieces of eyelash oil, use cotton swabs moistened with lens-cleaning fluid. For larger refractor lenses, use cotton balls dabbed in a solution of equal parts distilled water and isopropyl alcohol, with just a drop of mild dishwashing liquid.

Mirrors should be cleaned as seldom as possible since their delicate coatings are easily scratched. When it is necessary, use cotton balls soaked in distilled water and a drop of mild liquid soap. Rinse mirrors with pure distilled water, then let them stand on end so water spots don't dry on them.

Astrophotography

Astrophotography is one of the most challenging aspects of amateur astronomy. But the rewards—a meteor shower or the arcing trail of a comet , for example—make the effort worthwhile.

There are three ways the amateur can take astrophotographs: with a camera mounted on a tripod; with a camera and lens piggybacked on an equatorially mounted telescope; or with a camera hooked up directly to the telescope. The second and third methods require an excellent, rock-steady mount to track an object as it moves through the night sky—and considerable patience and skill.

For the beginner, it's best to start with a camera on a tripod. For the camera itself, the main requirement is a B (bulb) setting on the shutter that permits time exposures. The camera must allow the shutter to remain open for at least thirty seconds—preferably indefinitely. Avoid auto-exposure models with no override for aperture and shutter speeds. Consider buying an older camera with no automatic features.

Capturing a field of stars with the aperture set wide open is easy but requires fast lenses—from f/2.8 to f/1.2 is best. A basic set of lenses should include a wide-angle (24 mm or 28 mm), a standard (50 mm or 55 mm), and a short telephoto (85 mm to 135 mm). A 100-mm focal length is the minimum lens needed for taking pictures of the moon.

Only color film should be considered for astrophotography; black and white film offers no advantages, not even in cost. Fast films offer shorter exposures, but are grainier. Slow, fine-grained films record sharper detail than fast films. Use the slowest film that the subject allows. Film speeds are indicated by ISO numbers ranging from ISO 16 (the slowest fine-grained film) to ISO 3200.

The subject will determine the film speed. The moon is bright and detailed: ideal for ISO 100 film. But use at least ISO 400 film for

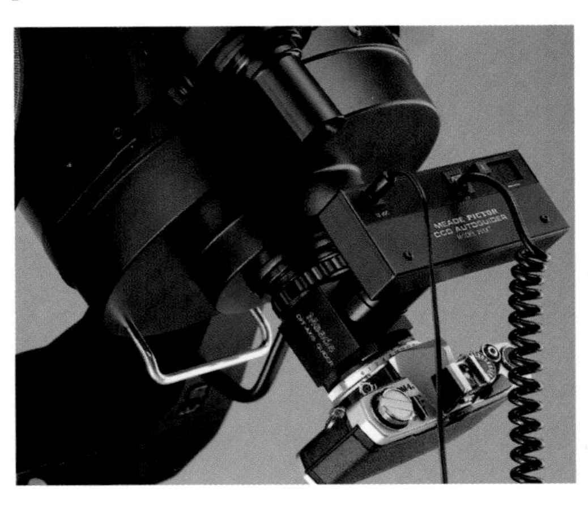

Advances in technology are improving the lot of the experienced astrophotographer. In the setup shown at right, a 35-mm camera is hooked up to a telescope equipped with an autoguider. The boxlike device is plugged into a drive unit on the telescope's mount and helps track stars precisely during long exposures—sometimes lasting more than an hour.

Thirty years ago, a shot like the one above of the Horsehead Nebula in Orion could only have been taken by a major observatory. Modern technology has changed that. This photograph was taken by amateur astronomer Jason Ware.

nightscapes with bright stars such as the Milky Way, and even faster film for meteors and comets.

The drill for a novice might go like this: Use a 35-mm camera with ISO 400 or faster film, a 55-mm or shorter focal-length lens, a tripod, and a cable shutter release to lock the shutter open for time exposures. Mount the camera on the tripod, turn off all automatic features, and set the lens to f/2.8 or the lowest f/ setting. Focus at infinity, set the shutter speed at B, and aim. Lock the mirror up if you can and press the cable release. Then take successive exposures of twelve, eighteen, and twenty-four seconds with a 50-mm or 55-mm lens; or twenty, thirty, and forty seconds with a 24-mm or 28-mm lens. Then release the cable lock.

THE DIGITAL REVOLUTION

Results once considered remarkable through photography are routine for digital imaging. With a CCD camera, film is replaced by a silicon chip called a charge-coupled device *(page 56)*. Much smaller than a frame of photographic film, CCD chips contain hundreds of thousands of tiny square pixels, or picture elements. A one-minute CCD exposure records about the same detail as a thirty-minute exposure on film.

This supersensitivity comes with its disadvantages. The main one is the cost of the CCD itself and the related expense of a computer and the necessary software. In addition, a superb mounting is a must. In general, leave this area of astrophotography to the intermediate and advanced amateur astronomer.

SECTION THREE

Observing the Nighttime Sky

UNDERSTANDING THE CELESTIAL SPHERE

The night sky may seem dauntingly mysterious, but its basic features are not difficult to understand. In many ways, our vantage point holds the key.

Anyone who has looked aloft on a summer evening knows the magic of the nighttime sky. But this glittering vista provokes endless questions even as it inspires wonder. Why are some stars bright and some just barely visible? Which ones are planets? Why do all the objects move across the sky in a certain way? How is it possible to find one particular object—say, a distant galaxy or a nebula—in the vastness of the night sky?

A time-exposure photograph centered on Polaris, the North Star, reveals concentric star trails, created as the Earth spins on its axis. Since Polaris is located above the Earth's northern axis of rotation, all the stars in the northern hemisphere appear to revolve around it.

First things first. The Earth is one of nine planets that orbit a star called the Sun, one of two hundred billion stars that constitute a disk-shaped, spiral galaxy—the Milky Way. The disk of our galaxy is four thousand light-years thick, and since we are embedded deep within it, there are stars extending about two thousand light-years above us and below us. These stars surround the Earth and fill the nighttime sky. During the day, however, the Sun's radiance overpowers the background stars and renders them invisible.

As the Earth orbits the Sun, it spins on its axis, so we spend half a rotation facing the Sun and half facing the night. This west-to-east rotation makes everything in the sky, including the Sun, appear to move from east to west. Times and angles of rising and setting vary according to the observer's location on Earth *(see below)*.

> *"The contemplation of celestial things will make a man both speak and think more sublimely and magnificently when he comes down to human affairs."*
>
> — Cicero
> 106-43 B.C.

WHAT'S UP THERE?

Under clear, dark skies, about three thousand stars are visible to the naked eye. All of them are located in the Sun's immediate vicinity within our galaxy. There are billions of stars farther away, along the disk of the Milky Way, but they can only be seen with a telescope.

That disk contains a host of fascinating deep-sky objects. There are star-forming nebulae, star clusters, and the remnants of novas—stars that suddenly blaze forth with increased brightness—and catastrophic stellar explosions, known as supernovas. Looking above and below the plane of the disk, we can find globular clusters—each one a spherical aggregation of a million stars, as well as other galaxies like our own that are millions of light-years away.

Planets orbit very close to stars while stars themselves are separated by light-years. Four planets are bright enough to see easily with the naked eye: Venus, Mars, Jupiter, and Saturn. Mercury and Uranus are harder to see, but are still visible without optical aid. Neptune appears like a small blue dot in binoculars, while Pluto looks like a very faint star even in large telescopes.

The Moon, our closest neighbor, orbits around Earth and appears as the brightest body in the nighttime sky. Its various phases can be seen at different times of the month. When full, however, its glare can obscure many of the fainter celestial objects.

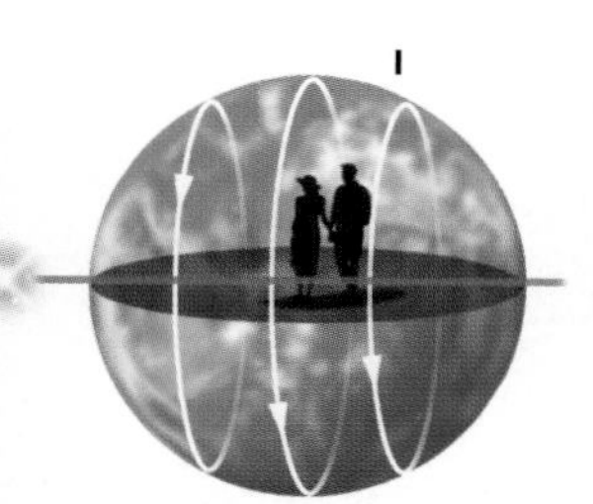

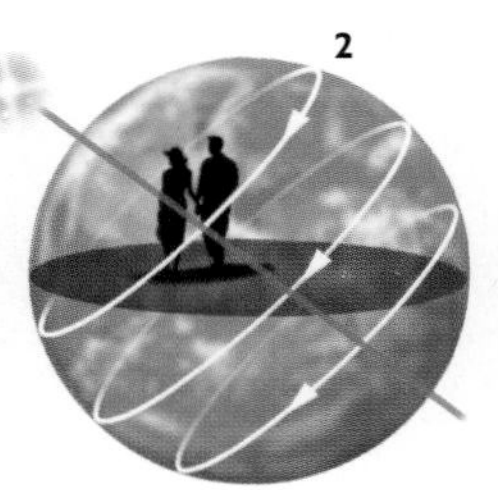

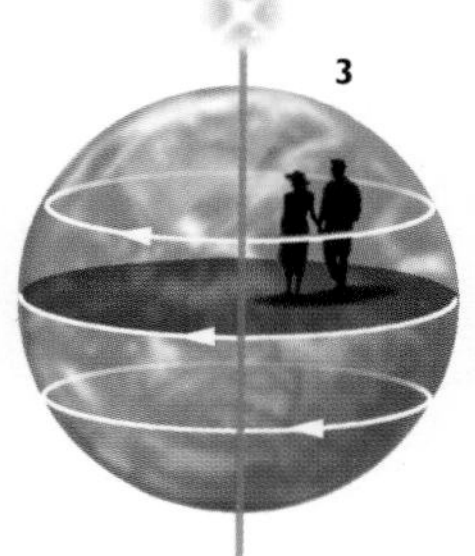

Different Locations, Different Motions
The motion of the objects in the night sky appears different depending on where an observer is standing on the globe. (1) At the equator, the stars rise and set vertically. (2) Between the equator and the north pole, the same objects rise and set at an angle, depending on the latitude. Observers in New York City at latitude 41°, for example, will see the stars rising and setting at a forty-nine-degree angle (90° - 41°) to the horizon. As the observer heads north or south, the stars' angle of rising and setting changes. (3) At the North Pole, the stars move horizontally across the sky. (In each example, the disk the observers are standing on represents their horizon.)

The Earth, Moon, and Sun

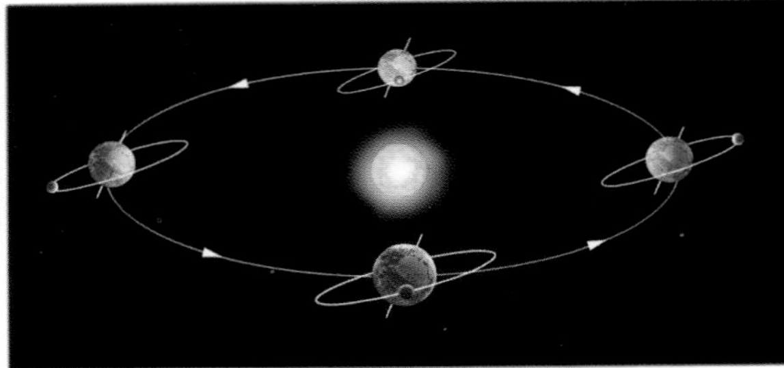

Seasons and Eclipses:
The tilt of the Earth's axis is responsible for the changing seasons (left). At the summer and winter solstices, the northern and southern hemispheres are tipped either toward, or away from the sun. Eclipses can only occur when the Earth, Sun, and Moon are in perfect alignment (below).

MOON
SOLAR ECLIPSE
MOON
LUNAR ECLIPSE

The gravitational interplay between the Earth, Moon, and Sun produces a cyclic pattern of orbital motion that is as constant as clockwork. This celestial dance is responsible for such phenomena as the seasons and the phases of the Moon. It causes eclipses, and it determines our cycles of night and day. In short, it governs life on our planet.

The Earth rushes through space at eighteen miles (30 km) per second in its annual journey around the Sun. At the same time, it also spins on its axis, completing one rotation each day. Meanwhile, the Moon circles the Earth at about twenty-three hundred miles (3,700 km) per hour, taking 27.3 days to complete one orbit. Viewed from above, all of these motions are counterclockwise.

The Moon's orbital motion can be seen as it advances by the equivalent of its own diameter every hour against the backdrop of stars. As the Moon orbits the Earth, the Sun illuminates it from different angles, which produces the lunar phases. The Moon also spins very slowly, completing one rotation every orbit, so that the same side of the lunar surface always faces the Earth.

The Earth's rotational axis is tilted 23.5 degrees from vertical, and since the Earth acts like a spinning gyroscope, its axis is always oriented toward the same point in space—almost. Actually, the Earth wobbles

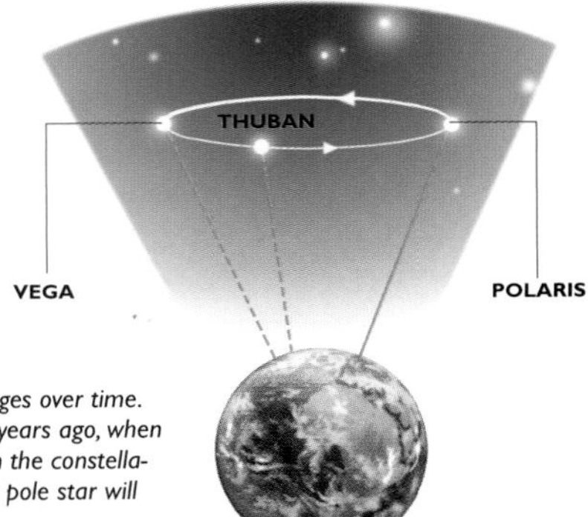

The Changing Pole Star
The Earth wobbles slightly on its axis, so its pole star changes over time. In the north, the current pole star is Polaris; five thousand years ago, when the pyramids were being built, the pole star was Thuban, in the constellation Draco. In another twelve thousand years the northern pole star will be Vega, in the constellation Lyra.

like a top, so its axis is not quite fixed. This wobble, called precession, has a 25,800-year cycle, which means that the Earth's north and south axes point to different pole stars over time *(opposite, bottom)*.

During part of its orbit, the northern half of the globe is tilted toward the Sun, while the southern half is tilted away: The points at which the Earth's axis reaches its maximum tilt are called the solstices. Viewed from mid-northern latitudes around the time of the summer solstice, the noonday Sun is at its maximum height, and its rays are direct and intense. The Sun also follows the longest arc across the sky at that time, producing long days and short nights. Six months later, at the time of the winter solstice, the Earth is on the other side of its orbit and the situation is reversed.

The two points midway between the solstices are called the vernal and autumnal equinoxes. At these times, the Earth tilts neither toward nor away from the Sun. Viewed from Earth, the noon Sun appears midway between its highest and lowest solstice positions. This produces days and nights of equal length.

1 NEW MOON

2 CRESCENT

3 FIRST QUARTER

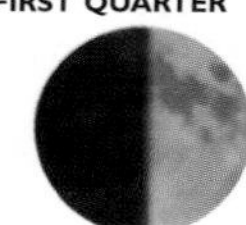

4 GIBBOUS

5 FULL MOON

6 GIBBOUS

7 THIRD QUARTER

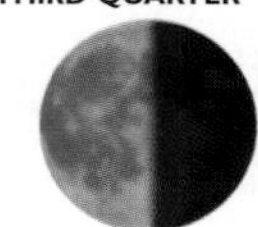

8 CRESCENT

WHY ECLIPSES OCCUR

Eclipses occur when the Earth, Moon, and Sun all line up. When the Moon passes directly between the Sun and Earth, we have a solar eclipse; or, it can pass through the shadow of the Earth, in which case the Earth causes a lunar eclipse *(opposite, top)*. Since the plane of the Moon's orbit around the Earth is tilted five degrees with respect to the Earth's orbit around the Sun, the Moon is usually above or below the "Earth-sun" line, which is why there aren't lunar eclipses each full Moon and solar eclipses every new Moon.

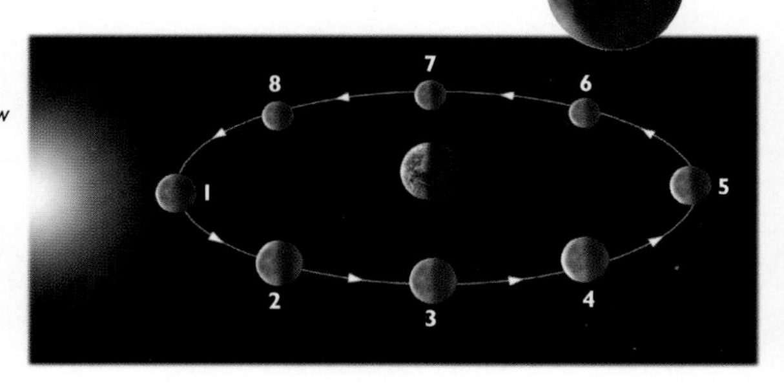

Lunar Phases

The Moon is fully illuminated (5) on the side of the Earth opposite the Sun. When the new Moon (1) and Sun are on the same side of the Earth the moon's Earthward face is in shadow. During the first- and third-quarter phases, the Moon is ninety degrees to the Earth and Sun and we see half of its face. About one week elapses between quarter phases.

The Planetary Ballet

The planets orbit the Sun in periods ranging from eighty-eight days for Mercury to 248 years for Pluto. Revolving around our star at varying distances, they all move at different speeds. The closer a body is to the Sun, the more rapid its travel along its orbit. As a result, the planets, as seen in our night sky, seem to move in an irregular way over the course of weeks or months. This seemingly random motion explains the choice of the word "planet," from the Greek for "wanderer." Understanding our place in the solar system helps to explain the movements we see.

The planetary ballet can be divided into two scenarios, depending on whether a planet's orbit lies within or beyond Earth's orbit. Mercury and Venus, the inner planets, never stray far from the Sun, as seen from our vantage point on Earth. Since their orbits occur within Earth's, they move from one side of the Sun to the other, appearing as morning stars, then evening ones. Emerging from a point of alignment with the Sun, known as conjunction, in the morning sky, they move rapidly westward at first, slowing as they reach a point called greatest elongation, where they appear farthest from the Sun *(below)*. Slowly at first, they seem to fall back toward the Sun, soon entering conjunction again and swinging out the other side into the evening sky. Interior planets can have a so-called inferior conjunction, when they lie between Earth and the Sun, or a superior conjunction, when they lie beyond the Sun. Outer planets can only have a superior conjunction.

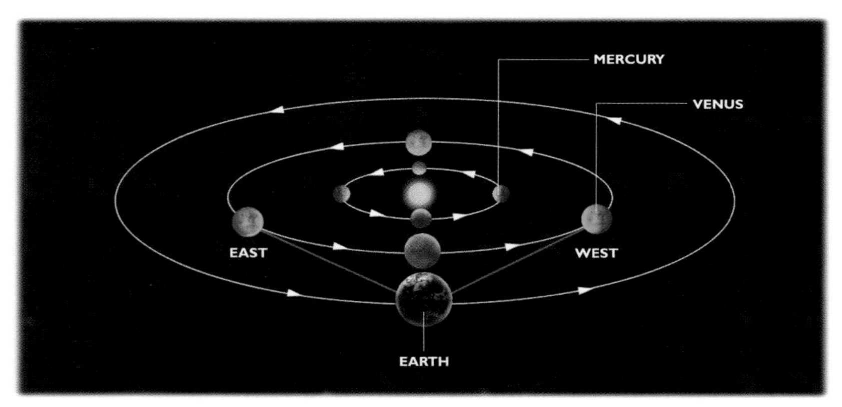

The Movement of the Inner Planets
The two planets with orbits that lie inside Earth's never stray far from the Sun. When at their farthest, the points are known as the greatest elongation west and east. Conjunction occurs when the Earth and Mercury or Venus are aligned with the Sun—inferior conjunction on the near side of the Sun and superior conjunction on the far side.

Mercury never appears farther than twenty-eight degrees from the Sun and is typically visible for only a few weeks during the course of a year. Venus wanders forty-seven

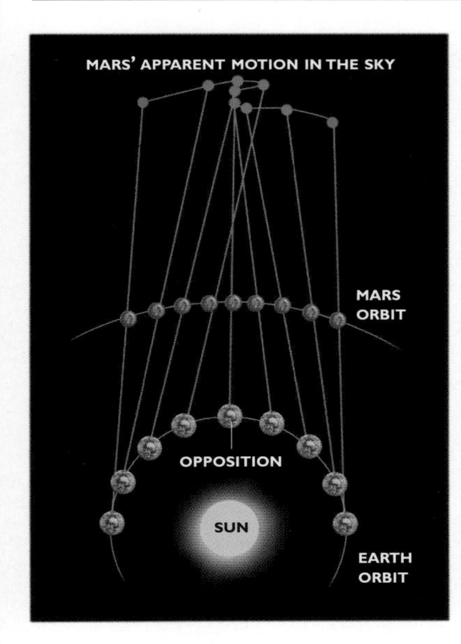

Retrograde Motion
When the orbit of the Earth overtakes a planet farther away from the Sun, the planet appears to move backward, then resume its forward motion.

degrees from the Sun and can appear as a prominent morning or evening star for many months on end.

THE RETROGRADE ILLUSION

The outer planets, by contrast, move across the sky in an easterly direction from year to year. The eastward motion seems to slow as the faster-moving Earth approaches an outer planet. Eventually the planet appears stationary in the sky, then begins a gradual westward, called retrograde, motion *(left)*. At the midpoint of retrograde motion, the Earth is at its closest point to the outer planet and an event called opposition is reached. The outer planet appears exactly opposite the Sun in the sky as seen from Earth and is visible all night. The retrograde period varies according to the planet's distance from Earth. Mars, for instance, spends two months of its 687-day orbit in retrograde motion, while distant Neptune spends most of the year.

As they move along their orbits, planets sometimes seem, as seen from Earth, to approach each other closely. This event, loosely known as a conjunction, is a line-of-sight alignment and not a near miss. Since all the planets, except Pluto, orbit near the ecliptic plane—Earth's orbital path around the Sun—conjunctions can sometimes be very close. And on rare occasions, planets pass in front of one another, an event known as an occultation. The next such occurrence will be in 2067, when Mercury occludes Neptune.

When the Earth and another planet are nearly lined up in their orbits around the Sun, they are said to be in conjunction. In this photo, Jupiter, Mars, and Mercury (near the top of the glow) can all be seen in close proximity to each other.

Coordinating the Sky

As the Earth spins on its axis from west to east, all the celestial objects appear to move in the opposite direction—from east to west each night. While the nocturnal parade of stars across the sky may seem confusing at first, celestial objects can be easily found and tracked by backyard astronomers with a few simple conventions.

Begin by visualizing the stars or planets as if they were painted on the inner surface of an immense, hollow celestial sphere. Astronomical objects can then be pinpointed on this dome with two intersecting coordinates. The two most commonly used methods are the horizontal or altazimuth system, based on the observer's horizon, and the equatorial system, based on the Earth's equator. Each has its advantages and disadvantages.

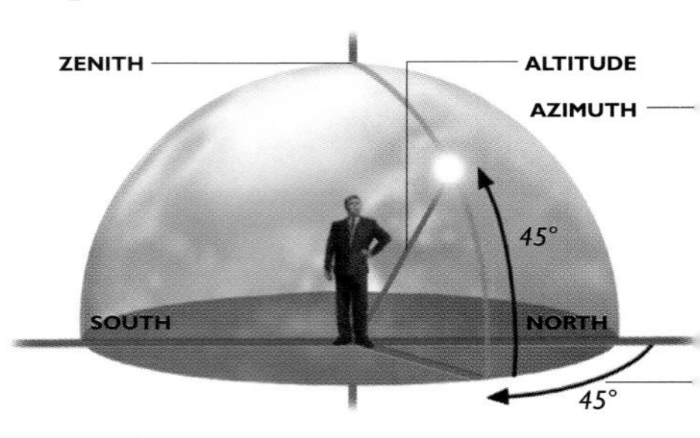

Locating Objects with Altazimuth
Altitude is the height of an object measured from the horizon to the zenith—the point overhead in the sky. Azimuth is the distance measured along the horizon, from true north, eastward to a point directly below the object. Here's how to find an object's altazimuth coordinates: From true north, scan eastward along the horizon until you reach the object's azimuth. From this point, scan upward to the object's altitude. So, an object located northeast halfway between the horizon and overhead would have an azimuth of 45° (the number of degrees from north counting eastward) and an altitude of 45°.

USING ALTAZIMUTH

The altazimuth system locates celestial objects using their altitude and azimuth, hence the name. Altitude is the height of an object measured from the horizon, 0°, to the zenith, the overhead point in the sky, which is 90°. Azimuth is the distance measured along the horizon from true north eastward to a point directly below the object *(above, right)*. True north is azimuth 0°, east is azimuth 90°, south is azimuth 180°, and west is azimuth 270°.

Although the altazimuth system is easy to use, it has limitations. For one, altazimuth coordinates are not

Basic Celestial Coordinates
A line extending through the Earth along its rotational axis is the polar axis, with a north and south celestial pole. The Earth's equator extended out to the sky defines the celestial equator. Zenith is the point directly above an observer. The opposite point is known as the nadir.

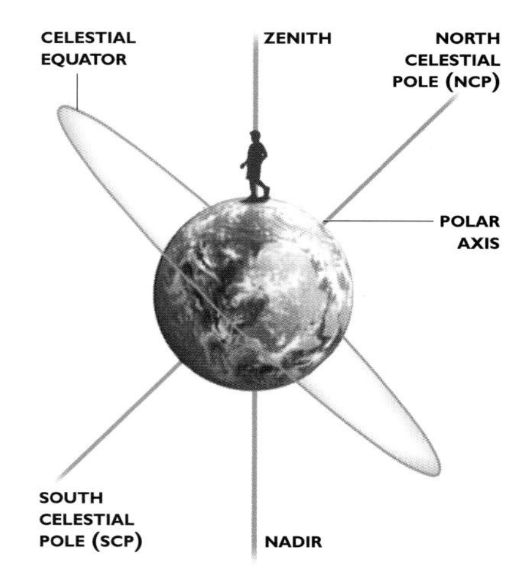

universal: They are based on the observer's horizon, and since they change with each moment, altitude and azimuth only apply to a given location at a given time. This makes it difficult to keep track of an object's coordinates and communicate them to other observers.

Another weakness of the altazimuth system is its inability to smoothly follow celestial objects across the sky. While the altazimuth system works perfectly at the poles, where celestial objects circle the sky parallel to the horizon, it is not suited to tracking the vertical motions of celestial objects seen from the equator or the diagonal motions seen elsewhere on Earth. This problem has been solved, however, with some altazimuth mounts on telescopes by the use of computerized controls *(page 73)*, which can make continuous horizontal and vertical adjustments.

THE EQUATORIAL SYSTEM

The equatorial coordinate system rectifies these shortcomings by simply extending the Earth's coordinates to the sky. In this system, the polar axes and equator are projected onto the celestial sphere, becoming the north and south celestial poles and the celestial equator. Likewise, terrestrial lines of latitude and longitude are projected onto the celestial sphere, becoming lines of declination and right ascension, respectively. These celestial lines are visualized as a gridwork that moves along with the sky. Each celestial object can then be pinpointed with respect to its "celestial longitude and latitude," or more appropriately, its right ascension and declination. Right ascension is abbreviated either as "RA" or with the Greek letter α, and declination is abbreviated either as "dec" or δ.

Since the equatorial coordinate system is oriented parallel to the terrestrial coordinates, it takes the Earth's rotation into account and allows celestial objects to be tracked smoothly with a single motion, without the aid of computers.

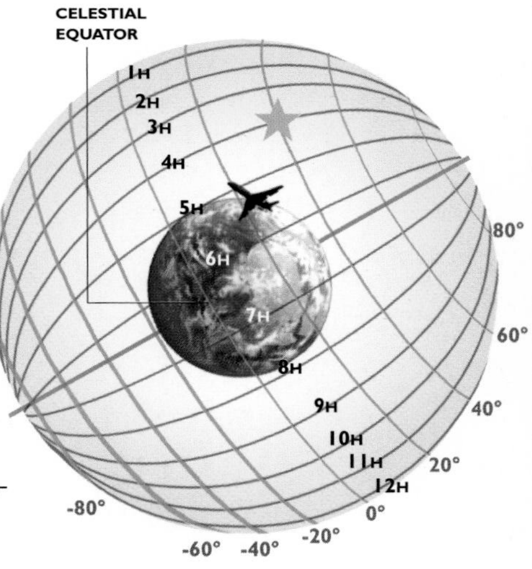

Equatorial Coordinates

Like terrestrial latitude, declination is measured in degrees. Declinations south of the celestial equator are designated with a minus (-) sign, while northern latitudes have no sign or a plus (+). The celestial equator is dec 0° and the north celestial pole is dec 90°. There are twenty-four-hour lines (h) of right ascension—celestial equivalent of longitude—that circumscribe the celestial sphere. They are counted eastward across the sky. Since Earth rotates 360 degrees in twenty-four hours, each hour of right ascension represents fifteen degrees of rotation. The equatorial coordinates of the star in this illustration are 4h, 40°.

A Matter of Time

For our earliest ancestors, time was simply the endless rhythm of nature. It was the flow of day and night; the changing of the seasons; the cycle of birth and death. As human intelligence grew, however, people viewed time differently: They began to measure it and make use of it.

Long before the planetary motions were understood, people counted the days, months, and years using the position of the Sun and the phases of the Moon. Days were measured from sunrise to sunrise; months were the number of days from full Moon to full Moon; and years were simply the number of days in the cycle of the seasons.

About six thousand years ago, Sumerians used sundials to divide the day into twelve equal parts; and they devised a calendar based on the 29.5-day cycle of lunar phases. About the same time, the Egyptians established the twenty-four-hour day and adopted a 365-day year.

Later, around 2000 B.C., the Babylonians subdivided the hour into sixty minutes, the minute into sixty seconds. These divisions were inspired by the 360-degree geometry

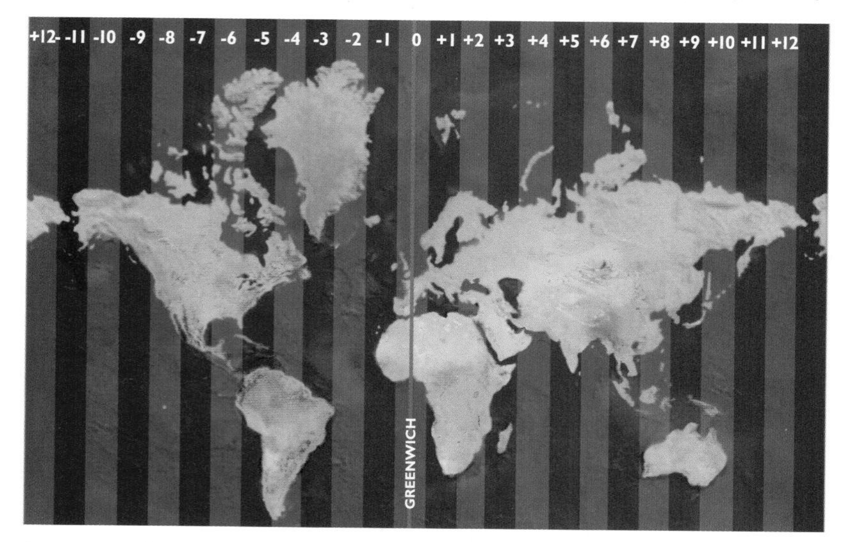

International Time Zones
Earth rotates fifteen degrees each hour and the twenty-four time zones reflect this fact. Successive time zones east of Greenwich (+) are an additional hour ahead; time zones west of Greenwich (-) are an additional hour behind. For example, Philadelphia's time zone is -5, which is five hours behind Greenwich. When clocks go ahead for daylight saving time, the time difference between Philadelphia and Greenwich becomes four hours instead of five.

Stonehenge in England was built in several stages between about 2800 B.C. and 1500 B.C. Archaeologists are still debating the purpose of Stonehenge, but it appears that the stone ring was used by its builders to predict solar and lunar eclipses, as well as mark the sunrise of the June solstice.

of the circle, which was founded on the Babylonian numerical system based on the number sixty.

The division of the year into 365 days worked well, except for the fact that a year does not have exactly 365 days. In 46 B.C. Julius Caesar revised the old Roman calendar and instituted the leap year, which kept everything on track—sort of. Although the Julian calendar, with its 365.25-day year, was an improvement, it was still slow by eleven minutes and fourteen seconds per year. The error slowly accumulated, and by 1582 the Julian calendar was ten days behind—so Pope Gregory XIII introduced some additional refinements. That year, the new Gregorian calendar skipped ahead ten days, taking up the slack.

NEW STANDARDS OF TIME

By the mid-1800s, global navigation made it necessary to establish an international time-keeping standard. Greenwich Mean Time (GMT) was initiated, based on the 0° line of longitude that passes through Greenwich, England. In 1884 the world was divided into twenty-four time zones, each one centered on the lines of longitude that run every fifteen degrees to the east and to the west of Greenwich. Finally, clocks everywhere could be coordinated with the local time at one spot on Earth—Greenwich.

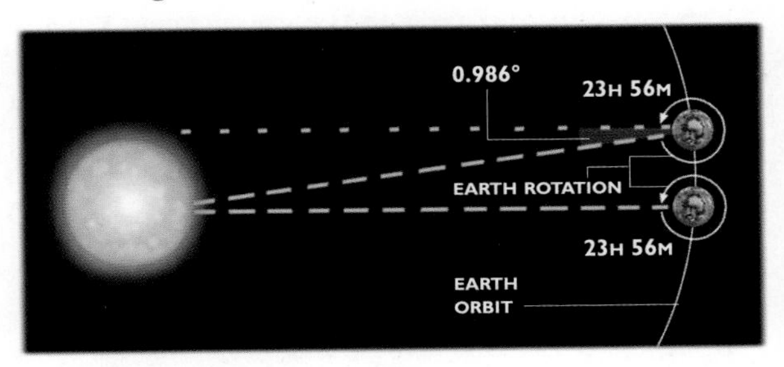

Sidereal Versus Solar Time
Earth's period of rotation is twenty-three hours, fifty-six minutes and four seconds, and it determines the rate at which the stars appear to move across the sky. Each day as the Earth advances around the sun, it must rotate an extra three minutes and fifty-six seconds—0.986 degrees— to bring the sun back to its same position in the sky. This additional period makes up the twenty-four-hour solar day.

In 1972 Greenwich Mean Time gave way to Universal Time (UT), which is virtually the same as GMT. Sidereal time, used by astronomers, is based on Earth's actual period of rotation—23 hours, 56 minutes, and 4 seconds—rather than the twenty-four-hour solar day.

GUIDE TO THE NIGHTTIME SKY

Since ancient times, humans in various civilizations have gazed up at the night sky and attempted to provide a convenient way to chart and classify what they observed.

In the same way that a country is made up of states or provinces, so the sky is divided into distinct areas, known as constellations. Every object in the sky lies within one of the eighty-eight constellations, which carve up the celestial dome into various irregular shapes with clearly defined boundaries.

"In her starry shade of dim and solitary loveliness, I learn the language of another world."

— GEORGE GORDON, LORD BYRON 1788-1824

Ancient and Modern Constellations
Ptolemy listed forty-eight constellations in the second century A.D., including Andromeda (above), *a northern constellation also known as Princess. Forty more "modern constellations" have been added since the 1600s.*

Stargazers in ancient times began naming constellations in honor of the gods, goddesses, and animals, the shapes of which they divined in the stars. Sumerians, who invented writing, were perhaps the first to organize stars into groups more than four thousand years ago, although civilizations such as the Chinese were also active in sky mapping. In the second century A.D., the Greek astronomer Ptolemy cataloged forty-eight constellations in his work *Almagest.*

Twelve of the best known form the zodiac and are part of astrological lore. The word zodiac means "belt of living things." Its importance in astronomy is due to the yearly path the Sun takes through the stars: the ecliptic. The paths of the Moon and planets also lie on or near the plane of the ecliptic and are thus part of the zodiac. The reason there are twelve constellations is that every time the Moon completes an orbit—27.3 days—the Earth advances one-twelfth the distance around the Sun, and the Sun appears to advance one constellation.

MODERN CONSTELLATIONS

Initially, the stargazers responsible for grouping patterns of stars lived in

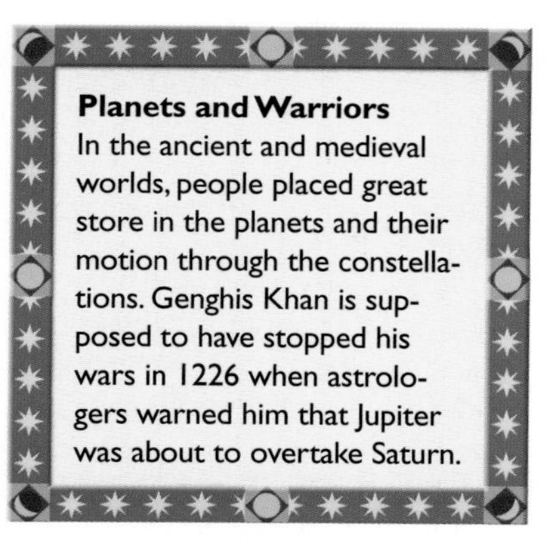

Planets and Warriors
In the ancient and medieval worlds, people placed great store in the planets and their motion through the constellations. Genghis Khan is supposed to have stopped his wars in 1226 when astrologers warned him that Jupiter was about to overtake Saturn.

the Middle East and Mediterranean area, and couldn't see parts of the southern sky. But sixteenth-century explorers sailing on southern seas began to remedy that situation. Many of the southern-sky constellations were named by the German astronomer Johann Bayer and by Nicholas de Lacaille, a French astronomer. Lacaille, somewhat partial to technical equipment, gave constellations such names as the Fornax (Chemical Furnace), the Horlogium (Clock), and the Telescopium (Telescope). However, constellations do not necessarily bear any resemblance to their names. For example, don't expect to find the microscope in Microscopium. It was named by de Lacaille to honor the inventor of the instrument and bears no resemblance whatsoever to the instrument itself.

With the telescope's introduction, more and more faint stars were discovered. But different astronomers continued grouping star patterns independently, resulting in many variations in atlases. In 1930 astronomers agreed to divide the sky into eighty-eight standard constellations that are universally used today. So, for example, an object like M57, the Ring Nebula, is referred to as the Ring Nebula in Lyra, the portion of sky defined by the constellation Lyra in which it is found.

Shifting Stars

It's certainly not going to happen overnight, but eventually, over a period of a hundred thousand years or so, the constellations we now see will change shape, just as they have in the previous hundred thousand years.

As illustrated by the Big Dipper at right (an example of an asterism—a recognizable pattern of stars), the stars slowly shift positions. This was first realized in 1718 by English astronomer Edmond Halley, when he noted that several bright stars had deviated from the positions recorded by the ancient Greeks. The movement appears extremely slow from Earth because of the huge distances that separate our planet from the stars.

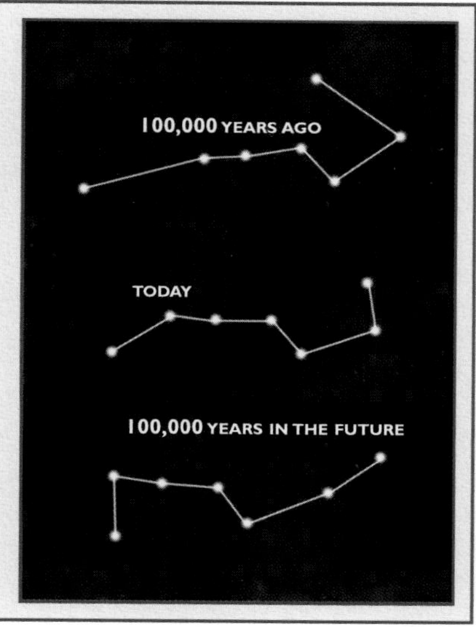

Mapping Stars and Constellations

Though novice stargazers have little difficulty in identifying the principal stars and constellations, even seasoned observers must rely on celestial catalogues and star maps to locate and identify the thousands of faint objects that populate the night sky.

Just as road maps help you find your way along highways and city streets, the star maps and constellation maps in the section that follows *(pages 100 to 175)* will guide you among a glittering celestial treasure of star clusters, nebulae and galaxies. But before you embark on your journey, you first need to become familiar with the basic tools for navigating the night sky.

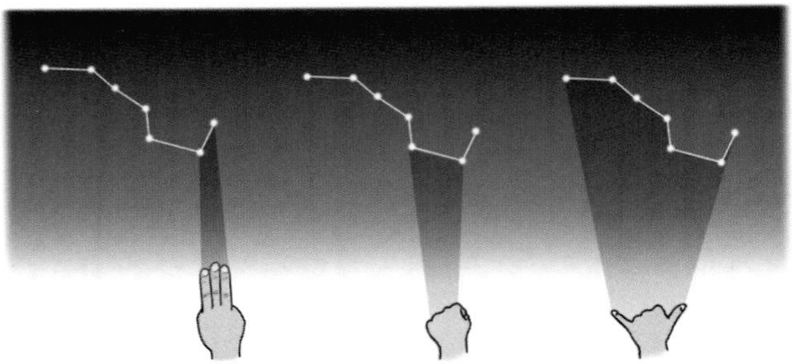

Handy Reference Guides

Distances between celestial objects are measured in degrees. The easiest way to judge these distances is with your outstretched hand: The width of your pinkie is roughly one degree; the middle three fingers make five degrees, the distance between the two pointer stars of the Big Dipper (above, left). *A fist corresponds to ten degrees, the width of the bottom of the Dipper's bowl* (above, center). *A spread hand will span the entire Dipper* (above, right), *the equivalent of twenty-five degrees. You can combine these measures to judge the size of a constellation. Andromeda, for example, is about fifty degrees wide, the equivalent of two spread hands, and thirty-five degrees high—a fist and a spread hand.*

ORGANIZING THE STARS

The stars differ in apparent brightness according to their actual luminosity and their distance from Earth. Since brightness is a star's most distinguishing feature, it forms the basis of stellar organization. In the second century B.C., the Greek astronomer Hipparchus divided the stars into six categories, designating the brightest as first magnitude and the faintest as sixth magnitude. The advent of the telescope early in the seventeenth century created the need for a more refined magnitude scale, improved star maps, and expanded catalogs of celestial objects and coordinates.

Though Hipparchus' magnitude system is still in use, it has been modified and extended: Today, a difference of one magnitude represents a brightness difference of about 2.5 times. For example, Sirius, the brightest star in the sky, has a magnitude of minus 1.46; Betelgeuse, a magnitude 0.50 star, appears 6.25 times fainter. The faintest stars visible to the naked eye are magnitude 6.0; the full Moon has a magnitude of minus 12.6; and the Sun, magnitude minus 26.8.

In 1603 the German astronomer Johann Bayer published an innovative celestial atlas entitled *Uranometria*, in which the principal stars of each

The Greek Alphabet

ALPHA	α
BETA	β
GAMMA	γ
DELTA	δ
EPSILON	ε
ZETA	ζ
ETA	η
THETA	ϑ
IOTA	ι
KAPPA	κ
LAMBDA	λ
MU	μ
NU	ν
XI	ξ
OMICRON	ο
PI	π
RHO	ρ
SIGMA	σ
TAU	τ
UPSILON	υ
PHI	φ
CHI	χ
PSI	Ψ
OMEGA	ω

constellation were given lower-case Greek letters *(left)*. In that system, α (alpha) was usually assigned to the brightest star, β (beta) to the second brightest, γ (gamma) to the third, and so on, although the star's position often played a role in the lettering sequence as well. Bayer's designations consist of the Greek letter followed by the constellation's genitive (possessive) form. Thus, alpha (α) Cygni is the brightest star in Cygnus, while alpha (α) Canis Majoris is the brightest star in Canis Major. Most of the brighter stars also have proper Arabic, Greek, or Latin names: For example, Vega, Arcturus, and Spica are the "alpha" stars of Lyra, Boötes, and Virgo.

When thousands of telescopic stars became visible, proper names and Greek letters proved insufficient. In the 1780s the French astronomer Joseph de Lalande numbered the stars in a contemporary edition of English astronomer John Flamsteed's *Atlas Coelestis*. Known as the Flamsteed number system, the stars of each constellation are numbered in ascending order from west to east. Hence, Betelgeuse is known as α Orionis and 58 Orionis—the fifty-eighth star from the western edge of Orion.

Sky Hopping
By locating Orion—a common sight in the winter sky—and using.your outstretched hand to judge degrees (opposite), it's relatively simple to hop around from one star to another. This commonly used technique enables newcomers to astronomy to find constellations that might otherwise be difficult to locate.

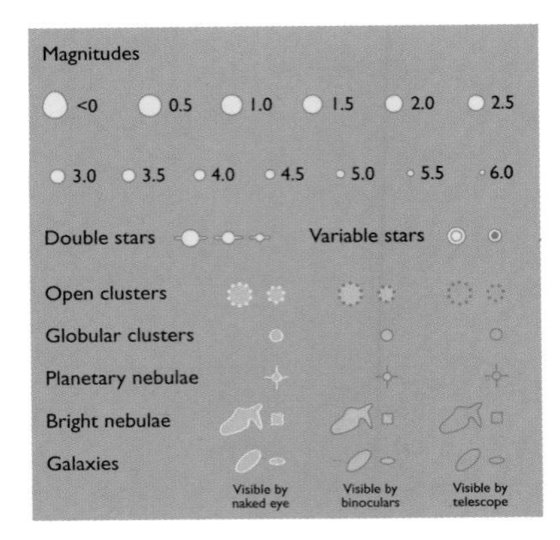

Symbols for Sky Charts
Both the bimonthly and constellation maps use these standard symbols to identify everything from star magnitudes to celestial objects.

CATALOGING THE SKY

By 1784, the noted French comet hunter Charles Messier had compiled a list of one hundred relatively bright, hazy objects that at first resembled tail-less comets but were clearly not. This list, originally intended to avoid confusion in the hunt for comets, eventually developed into the 110-object Messier Catalog known today. Messier objects are designated with an M followed by their number: They include thirty-nine galaxies, twenty-nine globular clusters, twenty-seven

Using the Bimonthly Maps
Choose the map that is designed for the month and time you are observing (pages 102-107). Next, determine your northern and southern horizons by drawing horizontal lines joining the latitude where you live, indicated along the left and right edge of the map. In the example at right, the horizon is set for an observer in Los Angeles at latitude 34 degrees. Hold the map overhead with the "North" side facing north. The point at the center of the map, midway between the horizon lines, is the zenith directly above you. In this case, the observer's zenith would be between Auriga and Perseus. Draco would be just above the northern horizon and Columba would be just above the southern horizon.

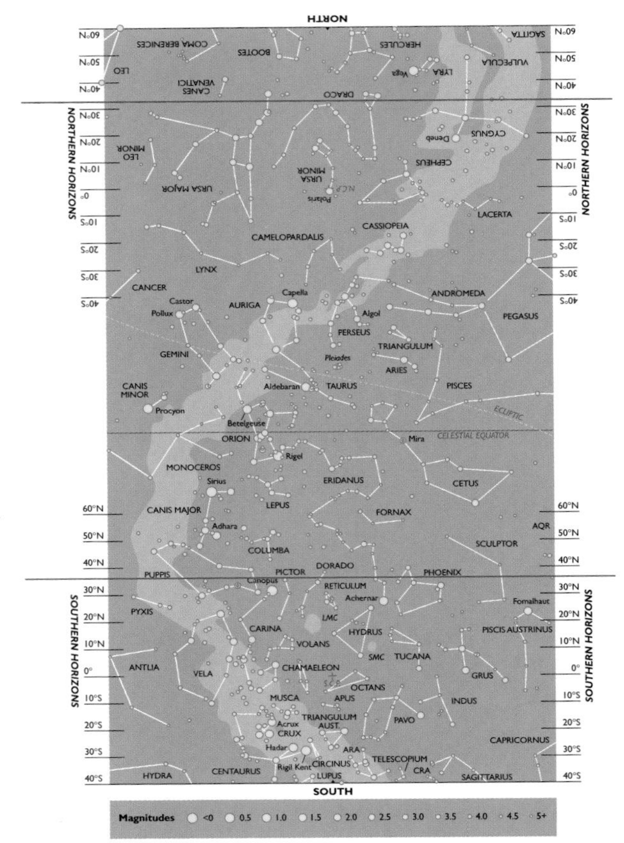

The name of the constellation, with a pronunciation guide and its common name.

Constellation area in square degrees, genitive pronunciation, best time and place for viewing, and size as gauged by hand.

CORVUS (KOR-vus) | *The Crow*

• **Abbreviation:** Crv • **Genitive:** Corvi • **Area:** 184 square degrees • **Size Ranking:** 70th

• **Best Viewed:** April to May • **Latitude:** 65ºN–90ºS • **Width:** • **Height:**

Most ancient civilizations depicted these stars as a bird. Greek mythology relates that Apollo fell in love with Coronis but was suspicious of her and sent a crow to spy on her. The crow told Apollo of her infidelity and for this service the bird was placed among the stars. The constellation is home to many galaxies, though most are not bright.

A brief description of the constellation, including relevant mythological lore.

In some cases, photos are included of more spectacular features.

NGC 4038/4039
The Ring-Tailed Galaxy is actually two galaxies that are interacting, connected by the force of gravity.

FEATURES OF INTEREST

DELTA (δ) CORVI: A bright double star, suitable for a small telescope. Component A is magnitude +2.95; component B is a dwarf star, magnitude +8.26. Current separation of the components is twenty-four arc seconds.

NGC 4038/4039: This is the notorious Ring-Tailed Galaxy, one of the finest examples of two galaxies in collision, although little evidence of the extraordinary processes taking place are visible with a small telescope. The northern component is the brighter of the two and small telescopes will show it as a peculiar double nebula, joined together in the east. Astronomers have long been aware of the so-called Antennae, thin sprays of matter being ejected as a result of the collision. Closeup views obtained by the Hubble Space Telescope show dark clouds of dust and gas signifying intense star formation.

A map of the constellation, including its boundaries in yellow and equatorial coordinates (page 89).

The key celestial objects visible in the constellation, with icons denoting whether they can be seen with the naked eye, binoculars, or a telescope.

open clusters, seven gaseous nebulae, four planetary nebulae, two asterisms (star groups), and one supernova remnant.

In 1888 Danish astronomer Johann Dreyer completed the exhaustive New General Catalog (NGC). The NGC listed 7,840 star clusters, nebulae and galaxies, although galaxies were not yet identified as such. Dreyer subsequently expanded the NGC by adding another 5,386 objects in two supplements called the Index Catalogs (IC), which were published in 1895 and 1908. Thus, the Orion nebula is known both as M42 and NGC 1976.

YOUR GUIDES TO THE STARS

Beginning on page 100, you will find two star maps that cover the entire northern and southern celestial hemispheres, followed by six bimonthly maps that show the changing sky over the course of a year *(pages 102 to 107)*. The bimonthly maps can be used as far north as Anchorage, Alaska, and as far south as Auckland, New Zealand.

Once you become more familiar with the general lay of the night sky, you can begin exploring in greater depth. The constellation maps *(pages 108 to 175)* give a detailed account of the many objects you will encounter on your celestial journey.

The Moon

Although the Moon is tantalizingly close to our planet, half of it is coyly concealed from Earth-bound observers. Because the Moon completes each rotation in exactly the same time it takes to revolve around the Earth, we can look at only one side of our nearest celestial neighbor. However, the so-called "far side" has been seen by orbiting spacecraft and a handful of astronauts who circled the Moon during the *Apollo* space program. Six of those missions landed men on the Moon between 1969 and 1972. Among the most prominent features, easily visible to the naked eye, are the maria—large, relatively smooth, flat plains. The word "maria" comes from the Latin word for seas, reflecting a mistaken belief by astronomers that these areas were large bodies of water. In fact, they are expanses of solidified lava that flowed from the interior of the Moon during its volcanically active period more than three billion years ago.

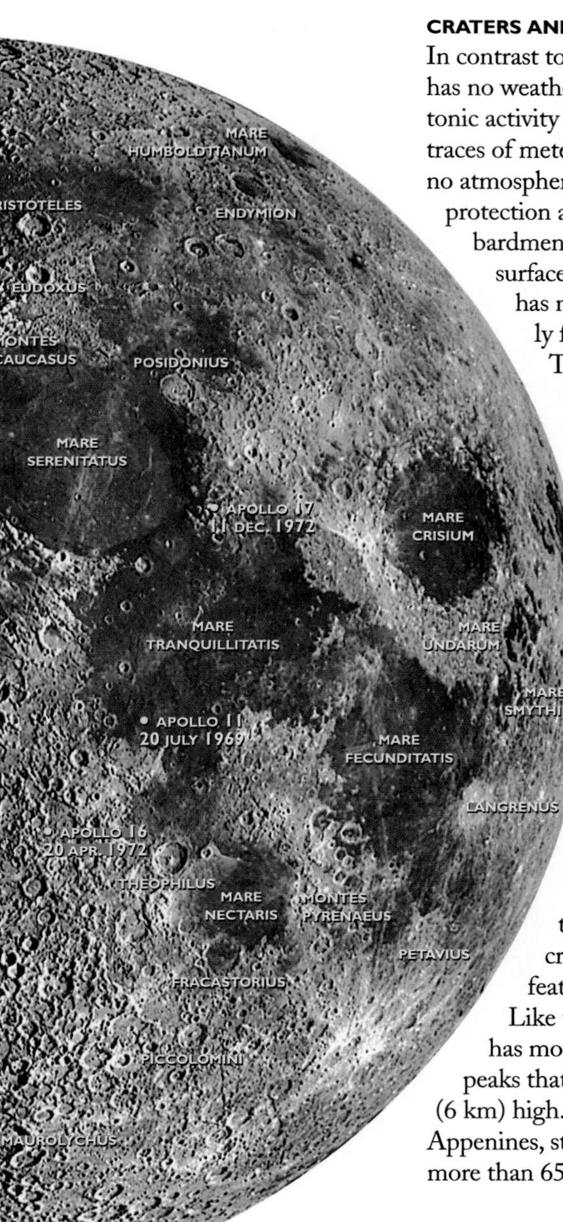

CRATERS AND RAYS

In contrast to the Earth, the Moon has no weathering and no plate tectonic activity that could erase any traces of meteor impact. It also has no atmosphere that could offer some protection against meteor bombardment. As a result, the lunar surface is heavily scarred and has not changed significantly for three billion years. The impacts of these meteorites have created a continuous range of impact features, from tiny "zap pits" to the far side's south-pole Aitken Basin, sixteen hundred miles (2,600 km) in diameter and more than seven miles (11 km) deep. When meteorites strike the lunar surface, they eject material from the impact site that spreads out radially, creating light-colored features known as rays.

Like the Earth, the Moon has mountains (montes), with peaks that soar up to four miles (6 km) high. The longest range, the Appenines, stretches a distance of more than 650 miles (1,050 km).

Sky and Constellation Maps

Before you take the sky and constellation maps outdoors, read the preceding section on Mapping Stars and Constellations *(pages 94 to 97)*. There, you'll find practical tips on how to navigate your way among the stars, as well as information on what the various maps contain.

This section begins with a map of the entire night sky for both northern and southern hemispheres *(below and opposite)*. The bimonthly sky maps on pages 102 to 107 show the sky as it appears throughout the year. The times are indicated at the top of the map. They are all listed in

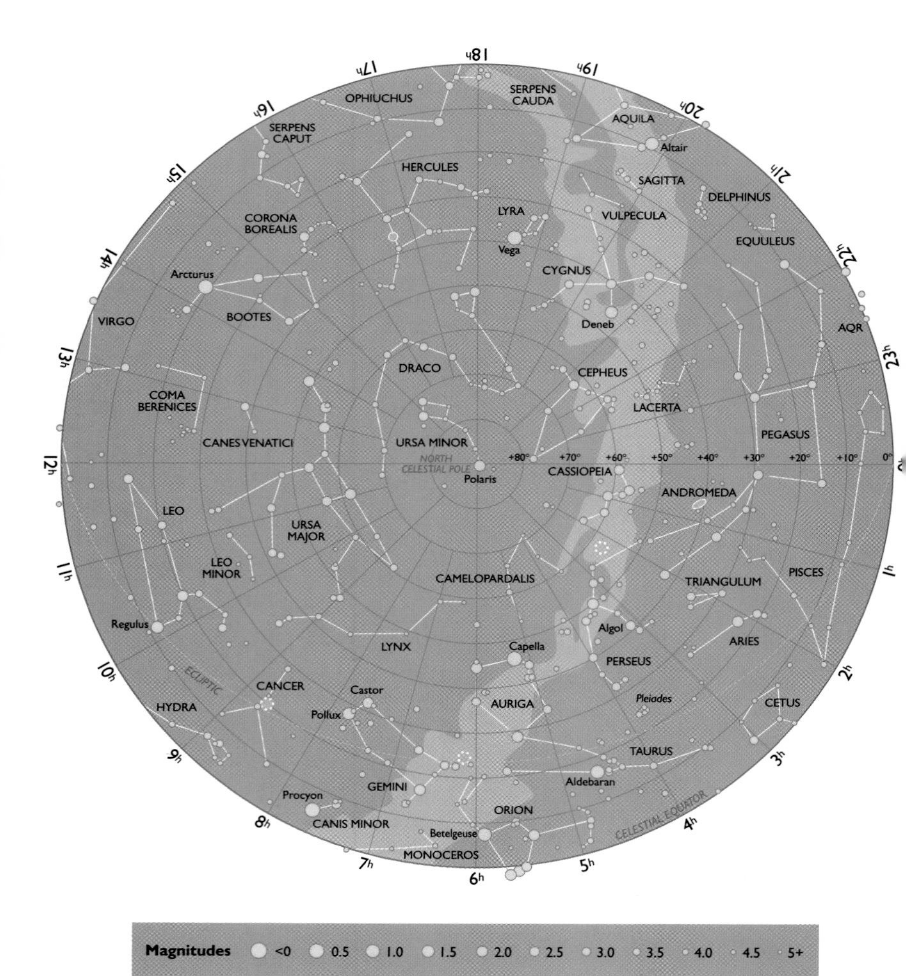

Standard Time and need to be corrected for Daylight Saving Time wherever applicable.

The ecliptic, the plane of the Earth's orbit around the sun, is marked with a yellow line. The wavy, light-colored band that passes through the maps is the Milky Way. The constellation maps on pages 108 to 176 provide lists of the objects that are to be found in a particular area of the sky. But first you will need to get your bearings. For that, all you require are the sky maps and a pair of binoculars, or just your naked eye.

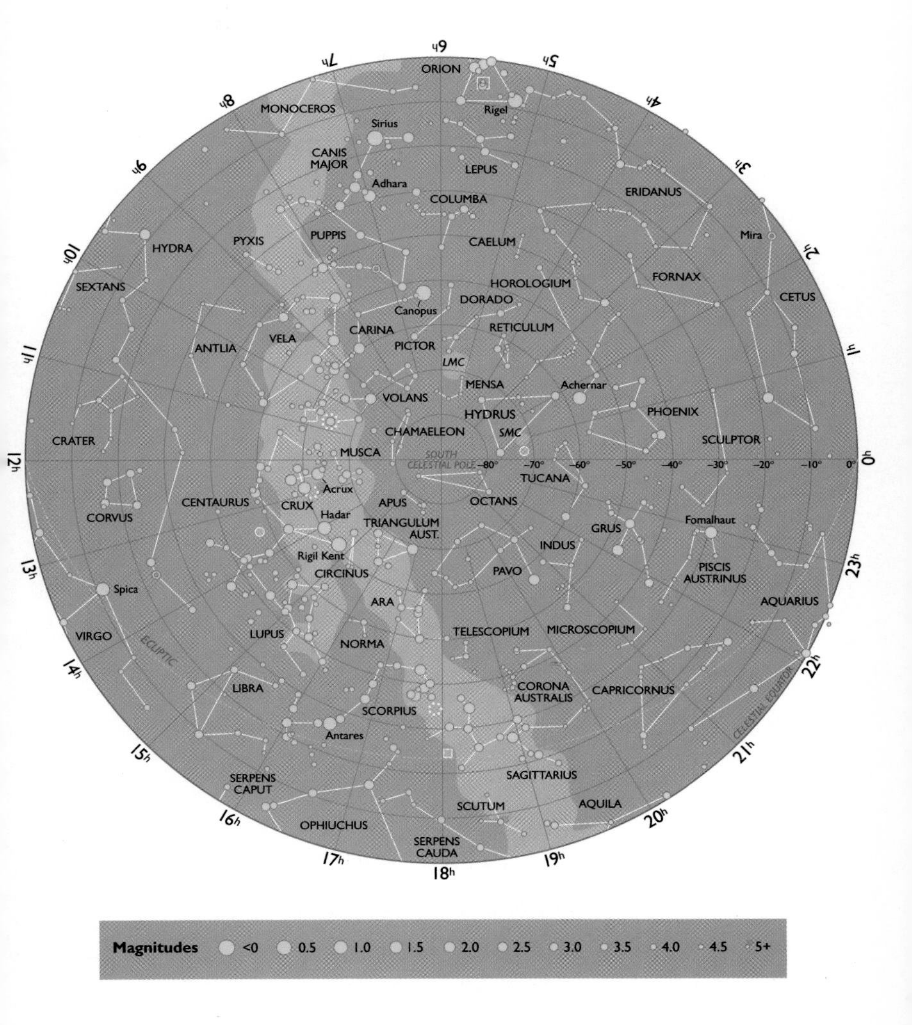

Sky map (a)

Late February 6 p.m.
Early February 7 p.m.
Late January 8 p.m.

Early January 9 p.m.
Late December 10 p.m.
Early December 11 p.m.

Late November Midnight
Early November 1 a.m.
Late October 2 a.m.

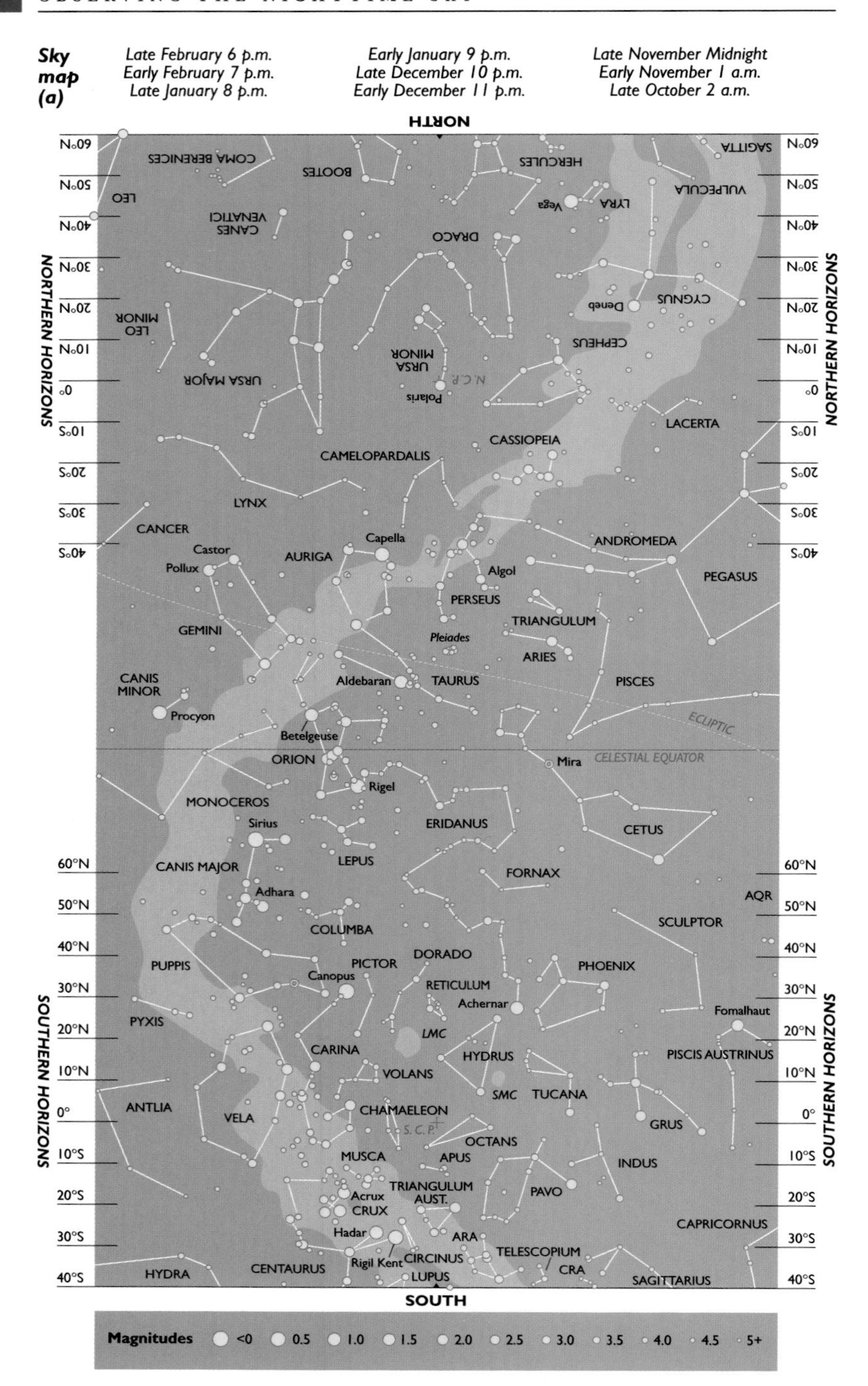

Sky map (b)

Late April 6 p.m.	Early March 9 p.m.	Late January Midnight
Early April 7 p.m.	Late February 10 p.m.	Early January 1 a.m
Late March 8 p.m.	Early February 11 p.m.	Late December 2 a.m.

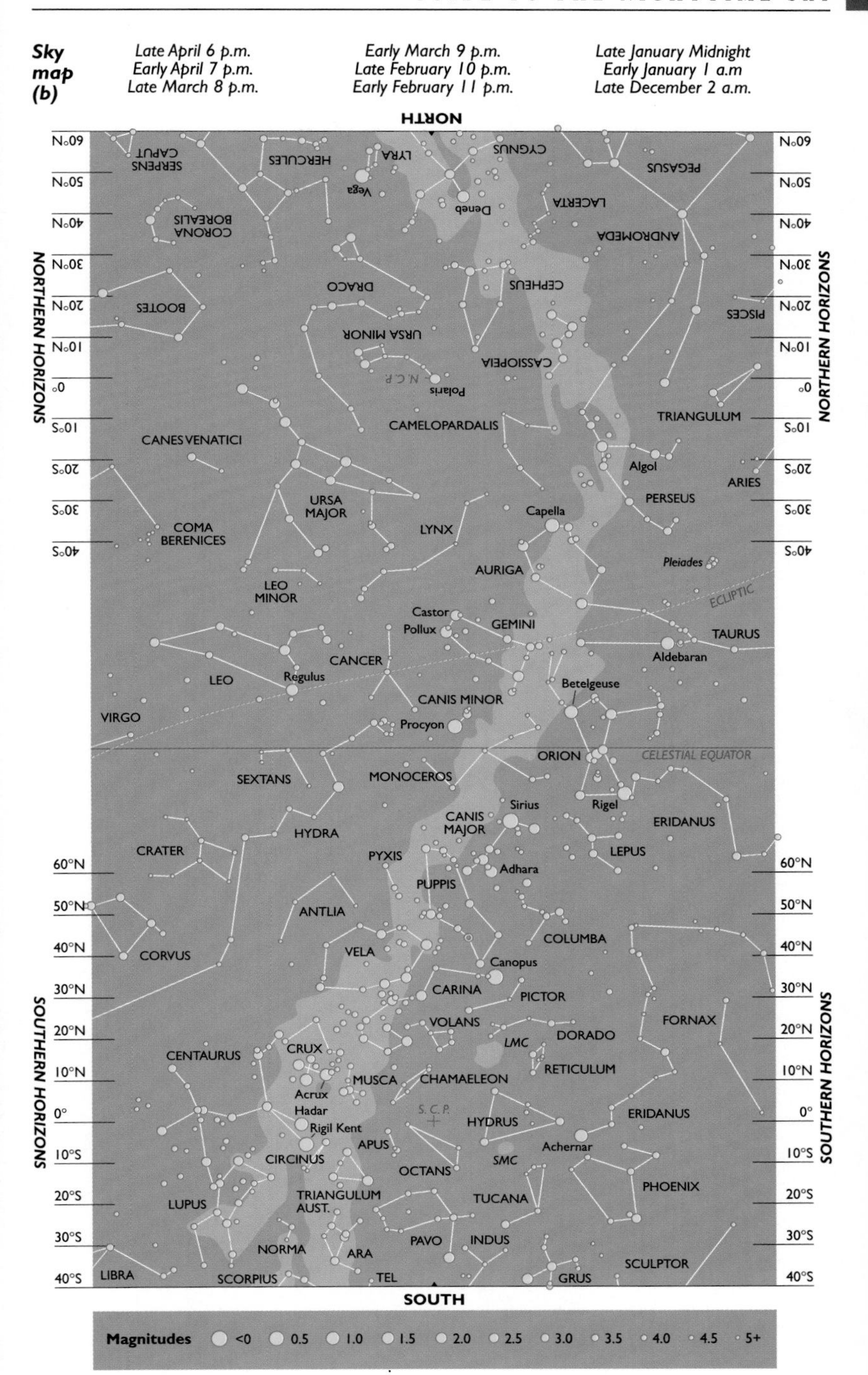

Sky map (c)

Late June 6 p.m.
Early June 7 p.m.
Late May 8 p.m.

Early May 9 p.m.
Late April 10 p.m.
Early April 11 p.m.

Late March Midnight
Early March 1 a.m.
Late February 2 a.m.

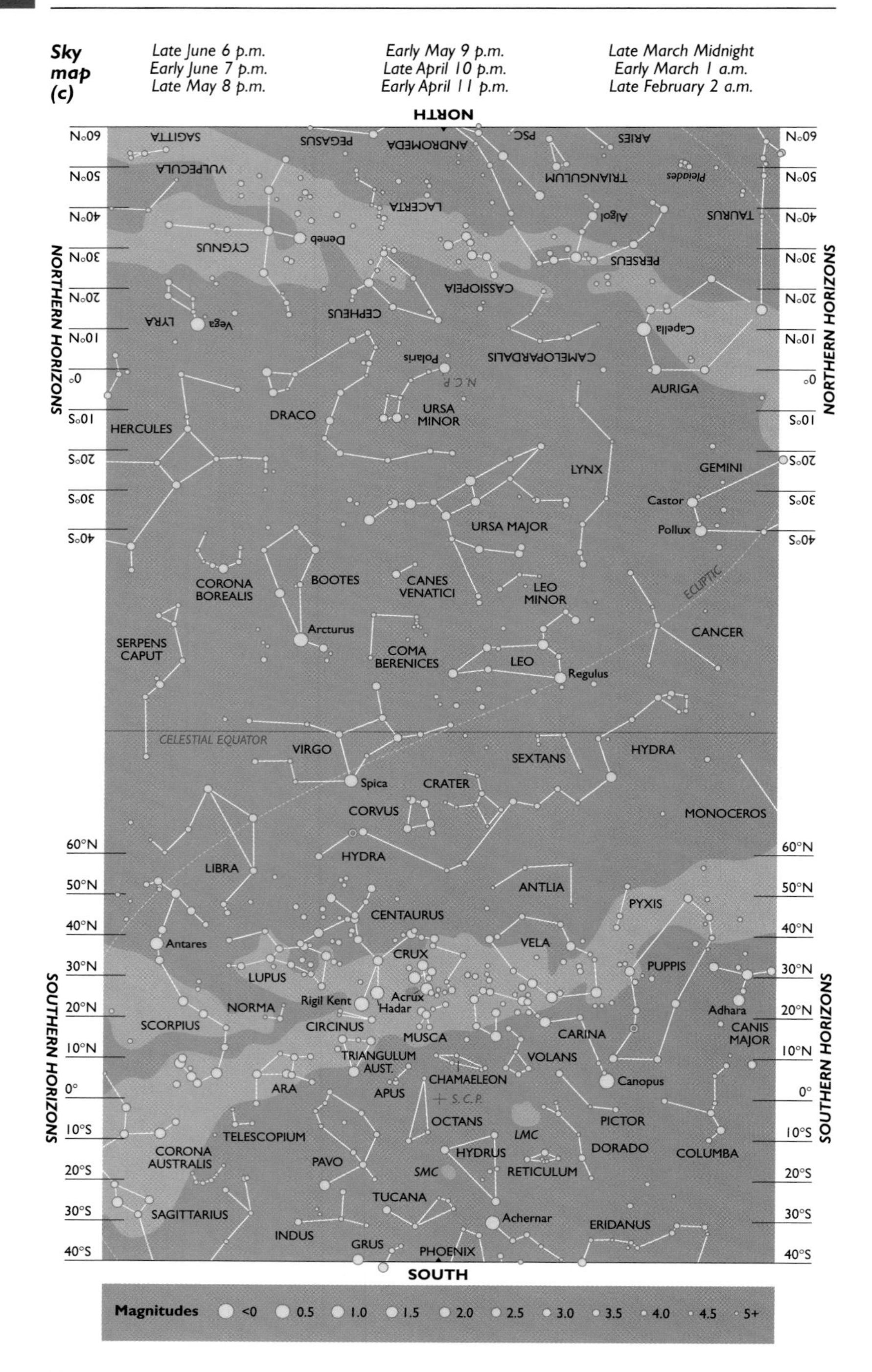

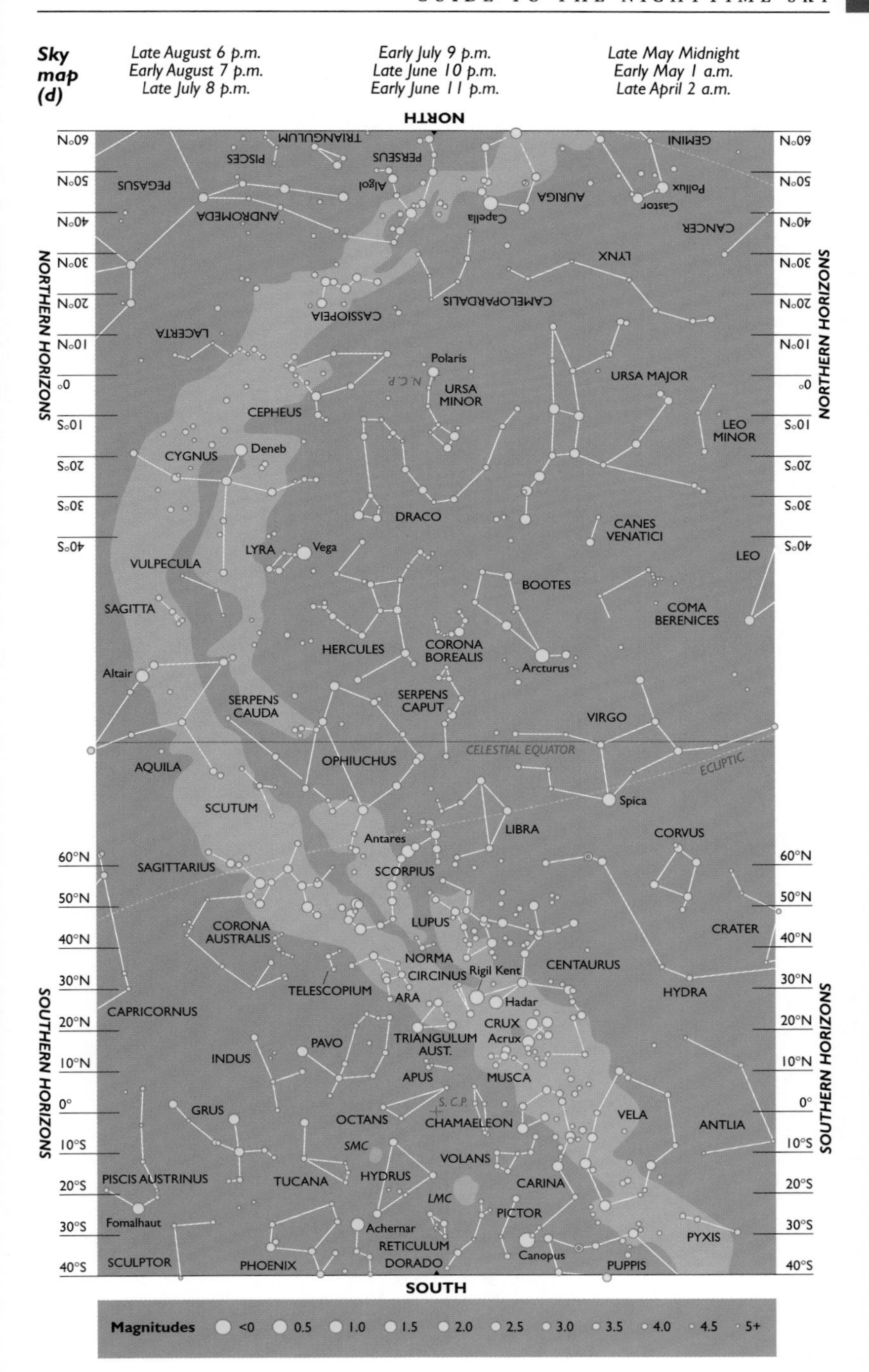
Sky map (d)
Late August 6 p.m.
Early August 7 p.m.
Late July 8 p.m.
Early July 9 p.m.
Late June 10 p.m.
Early June 11 p.m.
Late May Midnight
Early May 1 a.m.
Late April 2 a.m.
NORTH
SOUTH
NORTHERN HORIZONS
SOUTHERN HORIZONS
CELESTIAL EQUATOR
ECLIPTIC
PEGASUS
PISCES
TRIANGULUM
PERSEUS
Algol
ANDROMEDA
Capella
AURIGA
GEMINI
Castor
Pollux
CANCER
LYNX
CASSIOPEIA
CAMELOPARDALIS
LACERTA
Polaris
URSA MINOR
N.C.P.
URSA MAJOR
CEPHEUS
LEO MINOR
CYGNUS
Deneb
DRACO
CANES VENATICI
LYRA
Vega
LEO
VULPECULA
BOOTES
SAGITTA
COMA BERENICES
HERCULES
CORONA BOREALIS
Arcturus
Altair
SERPENS CAUDA
SERPENS CAPUT
VIRGO
AQUILA
OPHIUCHUS
SCUTUM
Spica
LIBRA
Antares
CORVUS
SAGITTARIUS
SCORPIUS
LUPUS
CORONA AUSTRALIS
CRATER
NORMA
CIRCINUS
Rigil Kent
CENTAURUS
TELESCOPIUM
ARA
Hadar
HYDRA
CAPRICORNUS
CRUX
Acrux
PAVO
TRIANGULUM AUST.
INDUS
APUS
MUSCA
S.C.P.
GRUS
OCTANS
CHAMAELEON
VELA
ANTLIA
SMC
VOLANS
PISCIS AUSTRINUS
TUCANA
HYDRUS
CARINA
LMC
Fomalhaut
PICTOR
Achernar
RETICULUM
PYXIS
SCULPTOR
PHOENIX
DORADO
Canopus
PUPPIS
Magnitudes
<0
0.5
1.0
1.5
2.0
2.5
3.0
3.5
4.0
4.5
5+

Sky map (e)

Late October 6 p.m.	Early September 9 p.m.	Late July Midnight
Early October 7 p.m.	Late August 10 p.m.	Early July 1 a.m.
Late September 8 p.m.	Early August 11 p.m.	Late June 2 a.m.

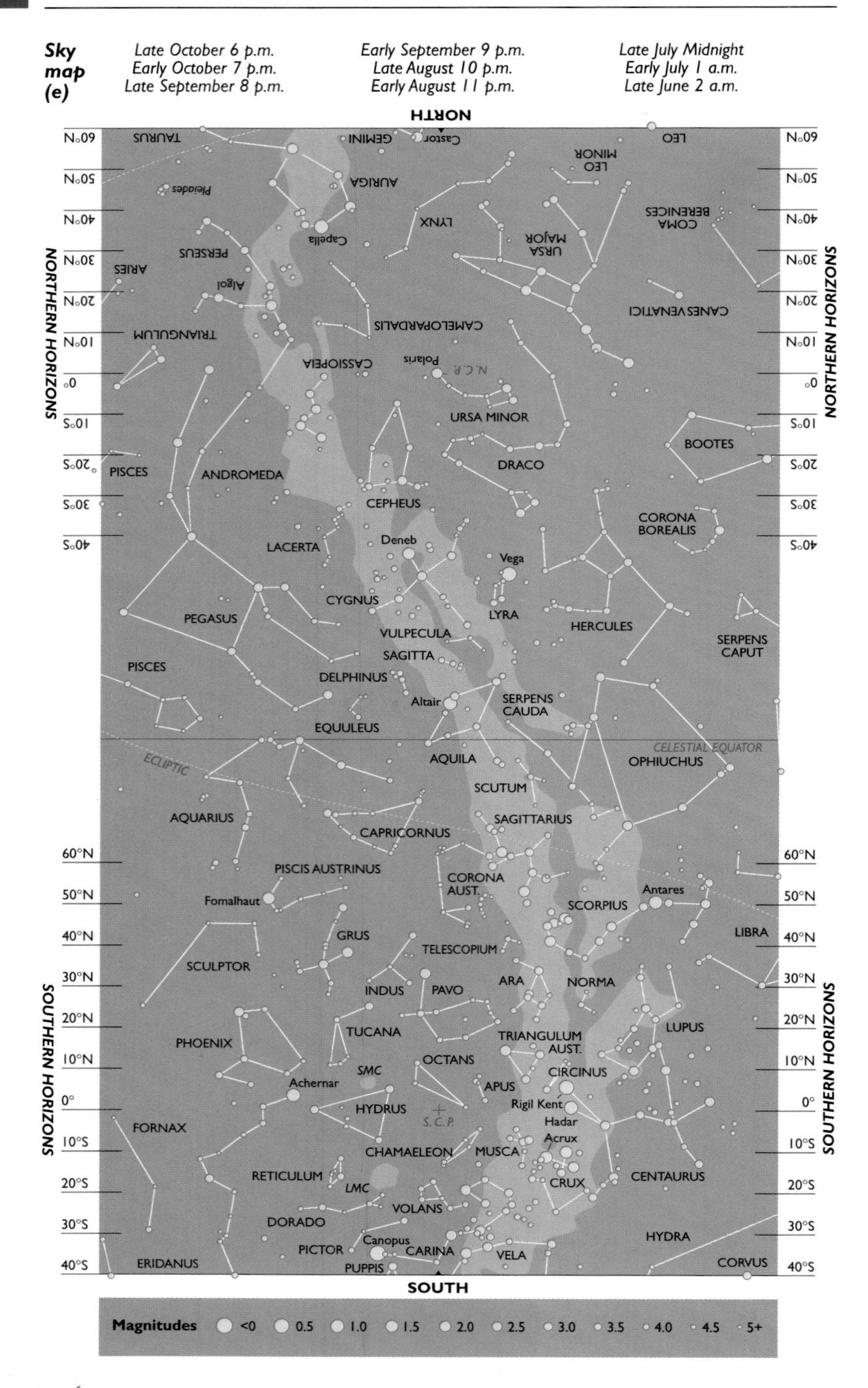

Sky map (f)

Late December 6 p.m.
Early December 7 p.m.
Late November 8 p.m.

Early November 9 p.m.
Late October 10 p.m.
Early October 11 p.m.

Late September Midnight
Early September 1 a.m.
Late August 2 a.m.

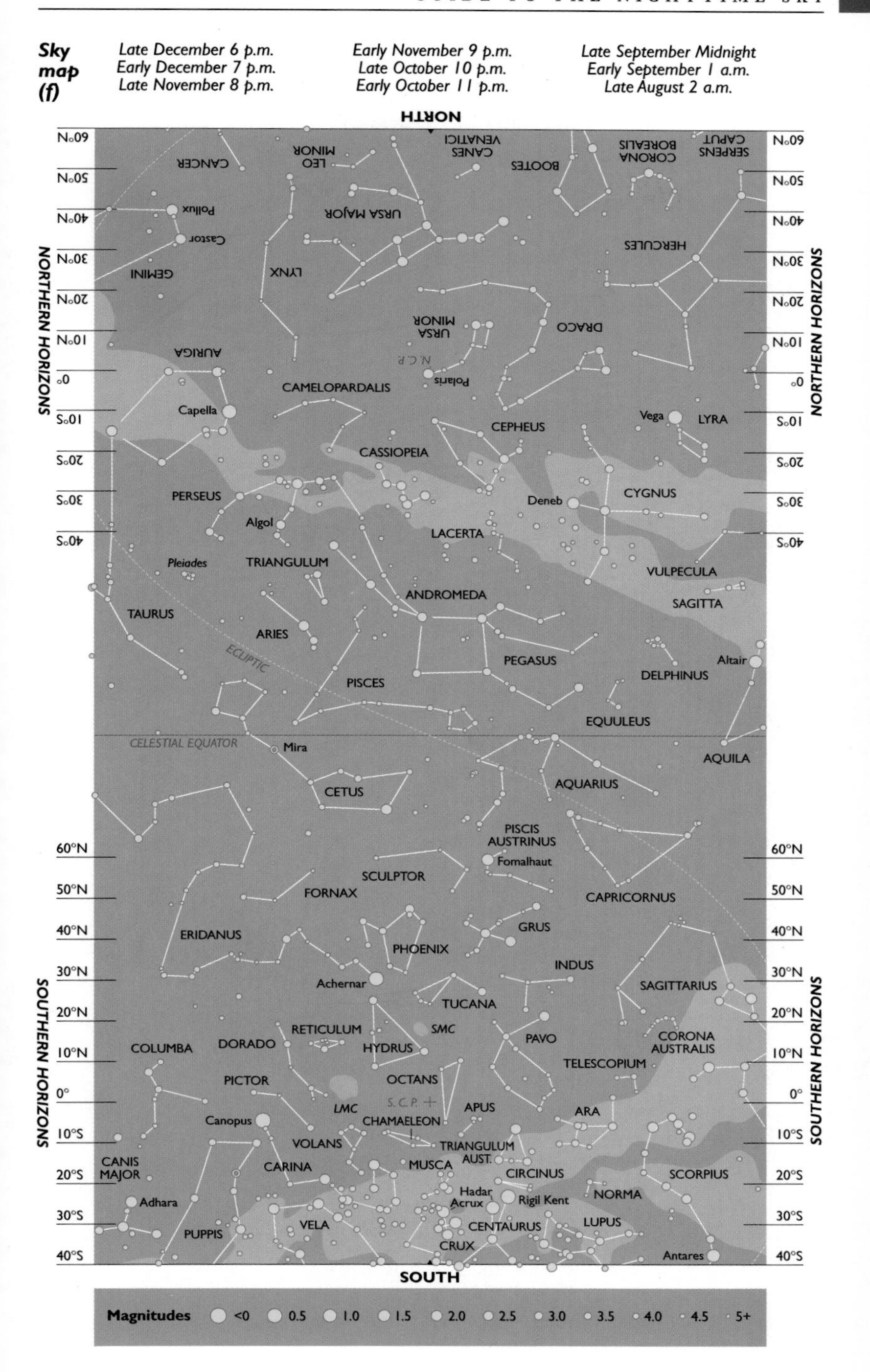

ANDROMEDA (AN-DROM-EH-DA) | *The Chained Princess*

• **Abbreviation:** And • **Genitive:** Andromedae • **Area:** 722 square degrees • **Size Ranking:** 19th

• **Best Viewed:** October to November • **Latitude:** 90ºN–40ºS • **Width:** 🖐🖐 • **Height:** 🖐✊

In Greek mythology, Andromeda was the daughter of King Cepheus and Queen Cassiopeia of Ethiopia. Chained to the rocks by her parents in the path of the sea monster Cetus, she was rescued by the hero Perseus, who slew the monster. This constellation shares its brightest star, Alpha (α) Andromedae, with the Great Square of the constellation Pegasus. The double row of second- to fourth-magnitude stars extending to the east is easily recognized in urban or rural settings and many interesting deep-sky objects can be found, owing to their proximity to bright members of the constellation. The constellation is home to a wide variety of objects: a bright star cluster suitable for binoculars, a major galaxy, an attractive binary star, and a bright planetary nebula.

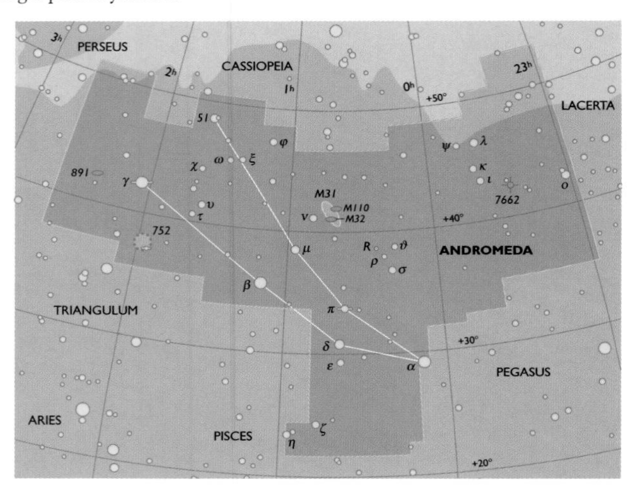

FEATURES OF INTEREST

BETA (β) ANDROMEDAE/NGC 404: If you've never seen a galaxy before, this is an easy way to find one: NGC 404, a magnitude +10.3 elliptical, is located only 6.4 arc minutes northwest of Beta Andromedae, so both objects can be observed at the same time. Easily visible with a six-inch (15-cm) telescope, this is a good test for the abilities of smaller instruments. The galaxy appears perfectly round with a bright core.

GAMMA (γ) ANDROMEDAE: This is a colorful double-star system for a three-inch telescope. The primary is magnitude +2.1 and yellowish while the secondary, located 9.6 arc seconds away, is blue and magnitude +4.8. The blue star has two companions, one of which is visible with very large amateur telescopes.

MESSIER 31 (THE ANDROMEDA GALAXY): At a distance of 2.2 million light-years, this galaxy is the most distant object visible to the unaided eye. M31 and our own Milky Way are the dominant members of the Local Group, a collection of twenty-five gravitationally bound galaxies (*page 9*). Most of the members are small, faint dwarf galaxies and because of their relative nearness, are located in all regions of the sky. The Andromeda Galaxy is thought to be larger than our own and has a population in excess of two hundred billions stars. M31 shows up well with 7x50 binoculars; the spiral arms can be traced for almost five degrees when conditions are right. The galaxy initially appears bright, though featureless, to beginners using a small telescope. With patience, however, something of its structure, including a major star cloud and at least two dust lanes, becomes visible. The core is very bright and brightens to a sharp stellar nucleus. Two dwarf companions can also be observed: Messier 32 is a small, round and very condensed elliptical galaxy located due south of M31's core, while Messier 110, slightly fainter but larger and more elongated, may be found northwest from M31. The entire field is very attractive, with countless faint stars, all members of our own galaxy, peppering the field.

M31

A spiral galaxy similar to our Milky Way, M31 (the Andromeda Galaxy) was once thought to be a nebula.

NGC 752: Located a little more than four degrees south-southwest from Gamma Andromedae, this large, coarse star cluster can easily be seen with binoculars. At least fifteen stars are brighter than magnitude +10 and the diameter of the cluster is about one degree. The cluster is best seen with a low-power eyepiece, which will provide a wide field of view.

NGC 891: Along with NGC 4565 in Coma Berenices, this is probably the best example of a spiral galaxy seen edge on, although it requires a fairly large aperture—at least eight inches (20 cm)—to be seen well. Its low surface brightness gives it a rather ghostly appearance. Telescopes with apertures of twelve inches (30 cm) and larger will show the dust lane that extends along the plane of the galaxy.

NGC 7662: This planetary nebula—a shell of gas surrounding a star late in its life—is also known as the Blue Snowball and is one of the dozen best objects of its class visible to the amateur astronomer. Though fairly small (with a diameter about thirty arc seconds) it is bright, and telescopes eight inches (20 cm) and larger will reveal a bluish tint.

M 32

A small, compact galaxy, M32 is a 9th-magnitude companion to its better-known neighbor, M31.

ANTLIA (ANT-LEE-UH) | *The Air Pump*

• **Abbreviation:** Ant • **Genitive:** Antliae • **Area:** 239 square degrees • **Size Ranking:** 62nd

• **Best Viewed:** March to April • **Latitude:** 45ºN–90ºS • **Width:** • **Height:**

This inconspicuous southern constellation, which represents an air pump, was named by the French astronomer Nicolas Louis de Lacaille in 1752. For the newcomer to observational astronomy, this is a rather disappointing area of the sky. There are many galaxies here that will interest amateurs who own moderately large telescopes, but only NGC 2997 is particularly bright. Binocular owners might find the faint naked-eye pair Zeta 1 and Zeta 2 interesting.

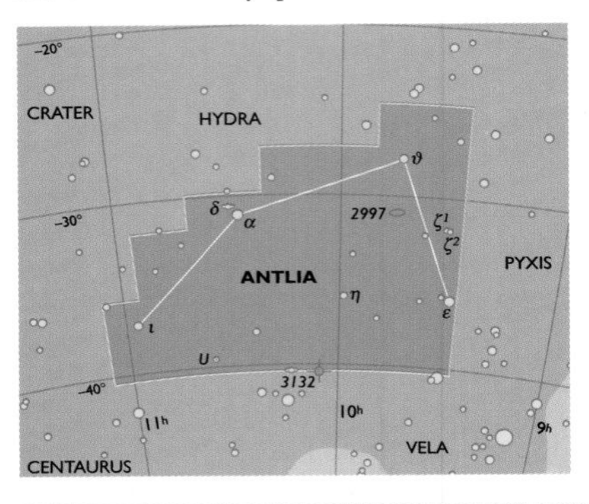

FEATURES OF INTEREST

ZETA (ζ) I ANTLIAE: This is a double star, magnitudes +6 and +6.5, separated by eight arc seconds.

NGC 2997: Spectacular in photographs, this galaxy appears as an oval haze with a brighter core, although rather dim with a three-inch (8-cm) telescope. Something of the spiral structure can be seen in 16-inch (40-cm) and larger telescopes.

APUS (AY-PUS) | *The Bird of Paradise*

• **Abbreviation:** Aps • **Genitive:** Apodis • **Area:** 206 square degrees • **Size Ranking:** 67th

• **Best Viewed:** July • **Latitude:** 5ºN–90ºS • **Width:** • **Height:**

The origins of this far-southern constellation are obscure, but it is thought to have been delineated by Johann Bayer in 1603. Bayer was the celestial cartographer who introduced the use of the Greek alphabet to classify the stars in all constellations in order of brightness, with alpha the brightest. The principal stars hover around magnitude +4.

FEATURES OF INTEREST

DELTA I (δ1) DELTA 2 (δ2) APODIS: This is a binary pair of stars wide enough apart to be seen separately with binoculars. Both stars are magnitude +5. They are separated by just under two arc minutes.

NGC 6101: This is a fairly typical globular cluster, a little fainter than average and rather loosely structured. Visible with three-inch (8-cm) and smaller telescopes, it requires about an eight-inch (20-cm) instrument to resolve into individual stars. NGC 6101 is a little less than fifty thousand light-years from Earth.

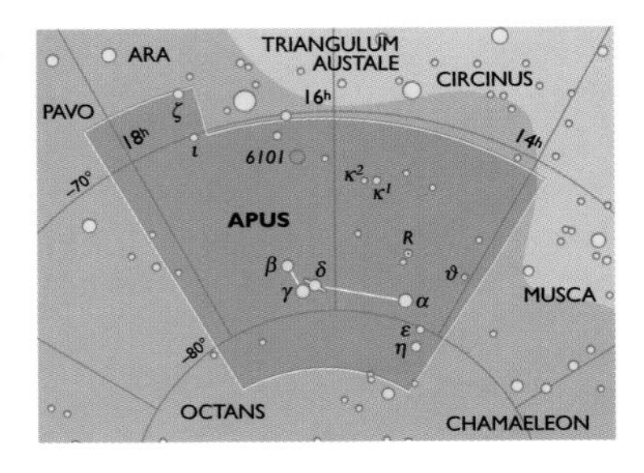

AQUARIUS (UH-KWAIR-EE-US) | *The Waterbearer*

• **Abbreviation:** Aqr • **Genitive:** Aquarii • **Area:** 980 square degrees • **Size Ranking:** 10th

• **Best Viewed:** August to October • **Latitude:** 65ºN–90ºS • **Width:** • **Height:**

The ancient Greeks, Arabs, and Babylonians all saw a man bearing jars or barrels of water in the stars of this constellation. Despite the fact that none of its stellar members is brighter than magnitude +2.93, this large constellation is quite distinctive and stands out well when observed from a rural location. Its most conspicuous feature is the Y-shaped asterism known as the Water Jar, located in the north-central part of the constellation. In addition to a number of bright, notable objects for the beginning observer, Aquarius is a rich hunting ground for faint, remote galaxies suitable for larger instruments, as well as one faint and seldom-observed globular cluster, NGC 7492.

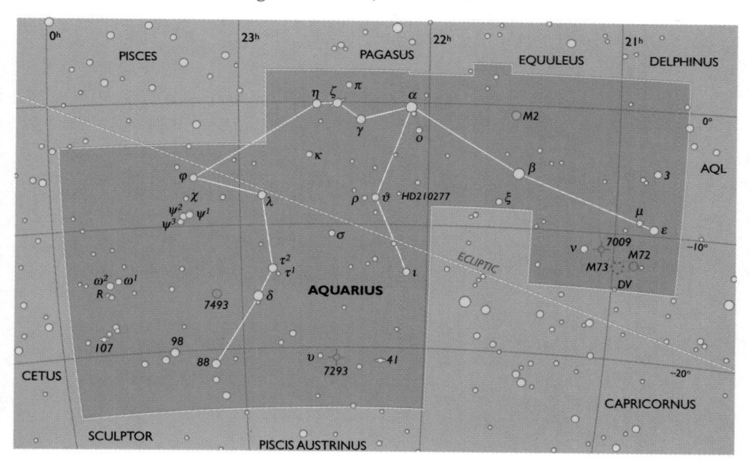

FEATURES OF INTEREST

ZETA (ζ) AQUARII: One of the stars that comprises the Water Jar, this double-star system has been followed closely since its discovery more than two hundred years ago. The components are magnitude +4.4 and +4.6. The separation between the two stars (two arc seconds) is slowly widening. The orbital period is believed to be 760 years.

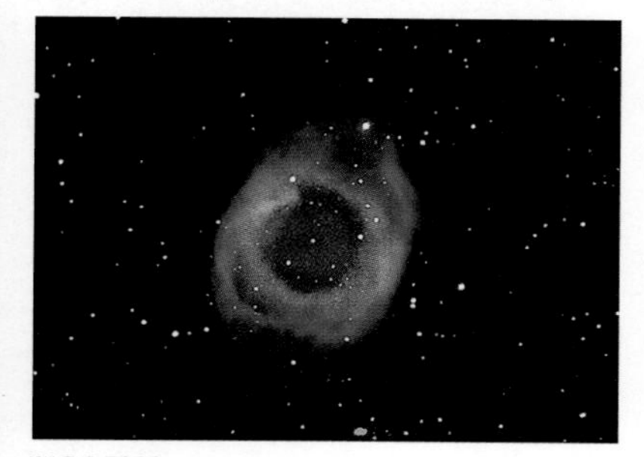

NGC 7293

Also called the Helix Nebula, NGC 7293 is an arresting sight in eight-inch (20-cm) and larger telescopes.

MESSIER 2: This is a very impressive globular cluster, detectable through binoculars and a pleasing sight with a small telescope. It shows a round, nebulous glow, quite bright and opaque, and brighter to the middle. About 6.8 arc minutes in diameter, the cluster is about forty thousand light-years distant.

NGC 7009: This bright planetary nebula was named the Saturn Nebula by Lord Rosse, who built a seventy-two-inch (183-cm) reflector in the nineteenth century. Rosse detected a thin line of faint nebulosity that reminded him of Saturn's rings, hence the name. These extensions are not visible with a small telescope, but the nebula itself, which has a mottled texture, is easily visible and quite well defined. Brightness: magnitude +8.3.

NGC 7293: This huge planetary nebula, called the Helix, is the closest object of its class to Earth, about five hundred light-years away. It is visible with binoculars as a faint, grayish glow about a third the diameter of the full moon. For good telescope views you will need at least an eight-inch (20-cm) lens with an interference filter to reveal the nebula's ringlike structure.

AQUILA (UH-KWIL-UH) | *The Eagle*

• **Abbreviation:** Aql • **Genitive:** Aquilae • **Area:** 652 square degrees • **Size Ranking:** 22th

• **Best Viewed:** July to August • **Latitude:** 85ºN–75ºS • **Width:** • **Height:**

The legends of all ancient middle-eastern civilizations associated this constellation with a giant eagle. The Greeks saw Aquila as the bird of Zeus, who carried Ganymede to the heavens, where he acted as the cupbearer of Zeus. This is a bright and easily identified constellation, dominated by first magnitude Altair. At a distance of 16.78 light-years, Altair is one of the sun's nearest neighbors and was used by *Apollo* astronauts to help navigate their spacecrafts. The Milky Way slices directly through the constellation, but despite this, Aquila is notably lacking in bright nebulae and interesting star clusters. There are, however, planetary nebulae and several absorption nebulae—cold dust clouds that absorb light from bright stars behind them.

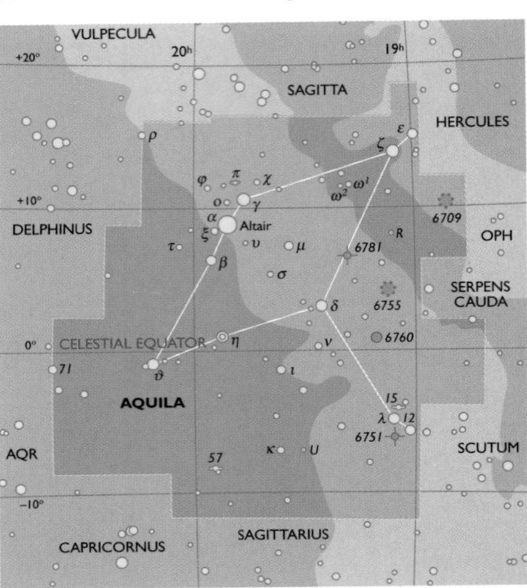

FEATURES OF INTEREST

15 AQUILAE AND 57 AQUILAE: Two double stars suitable for viewing by steadily mounted, high-powered binoculars.

ETA (η) AQUILAE: This variable star is a Cepheid variable, of which its prototype is the star Delta Cephei. These stars vary in brightness in regular, precise intervals, the result of pulsations of the star's atmosphere. Their luminosities are proportional to the length of their periods, so astronomers have used Cepheids, which are typically bright stars, as distance indicators to nearby galaxies. Eta Aquilae varies over a period of 7.176 days from a maximum of magnitude +3.7 to a minimum of +4.5.

NGC 6755: A gathering of magnitude +11 and fainter stars. The central region appears slightly nebulous. The surroundings are relatively starless so the cluster stands out well.

NGC 6760: In a field peppered with dim stars, this globular cluster appears as a faint haze slightly elongated in a north/south direction.

NGC 6751 AND NGC 6781: These two planetary nebulae are the best Aquila has to offer. NGC 6751 is rather small and faint in eight-inch (20-cm) and smaller instruments; the central star can just be detected at 100x magnification. NGC 6781 is much larger (about two arc minutes in diameter) though rather pale nebula, smooth textured, and brighter along its southern boundary.

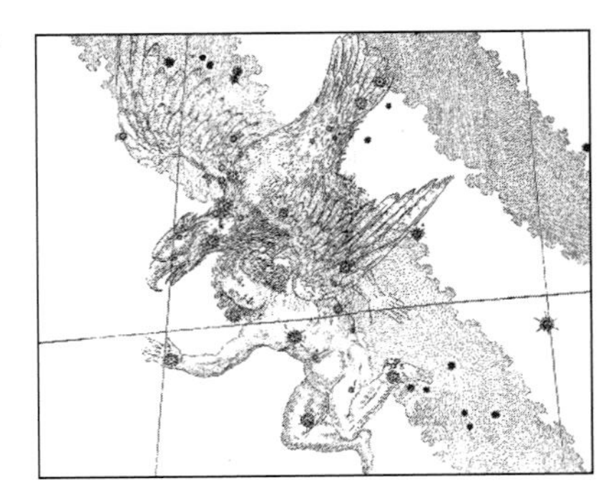

The stars in this sector of the sky have been identified as an eagle since before 1000 B.C. The ancient Greeks believed that the bird held the thunderbolts of Zeus.

ARA (AR-UH) | *The Altar*

• **Abbreviation:** Ara • **Genitive:** Arae • **Area:** 237 square degrees • **Size Ranking:** 63th

• **Best Viewed:** June to July • **Latitude:** 25ºN–90ºS • **Width:** • **Height:**

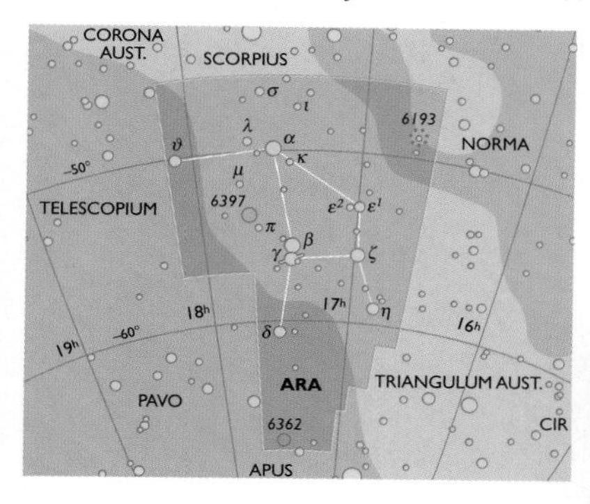

This southern constellation was one of Ptolemy's original forty-eight. Though small, it is nestled in the Milky Way just south of Scorpius and is made up primarily of second and third magnitude stars. Its brightest is Beta Arae (magnitude +2.87). For the ancient Greeks and Romans, the Altar was the throne of the centaur Chiron, considered the wisest being on Earth.

Before the creation of the modern constellation Norma, the stars of Ara were once part of Centaurus and Lupus.

FEATURES OF INTEREST

NGC 6193: This open star cluster is large and scattered over an area roughly half the size of the full moon. The brightest star is sixth magnitude.

NGC 6397: This bright globular cluster is a loose assemblage of stars for objects of this class. Not particularly massive, it appears rather large in small telescopes, owing to its proximity to Earth. At nine thousand light-years distance, NGC 6397 is the nearest globular cluster to our planet.

ARIES (AIR-EEZ) | *The Ram*

• **Abbreviation:** Ara • **Genitive:** Arae • **Area:** 237 square degrees • **Size Ranking:** 39th

• **Best Viewed:** November to December • **Latitude:** 85ºN–75ºS • **Width:** • **Height:**

In Greek mythology, Aries the Ram was clothed in golden fleece and was sent by Hermes to carry Phrixus and Helle, sons of the kings of Thessaly, away from their evil mother. Flying high over the ocean, Helle lost his grip on the fleece and fell to his death. When his brother reached safety, he sacrificed the ram to the gods in gratitude. This is a small constellation of the zodiac, but its three brightest stars, alpha (α), beta (β), and gamma (γ) are conspicuous, even in a light-polluted environment.

FEATURES OF INTEREST

GAMMA (γ) AREITIS: Identified by English scientist Robert Hooke in 1664, this is one of the first double stars discovered in the telescopic age. Both components are white and equal in brightness (magnitude +4.5), separated by 7.8 arc seconds.

LAMBDA (λ) AREITIS: Visible with tripod-mounted, high-power binoculars, this pair of stars are magnitudes +5 and +7.5.

NGC 772: This is the brightest galaxy in Aries and detectable with a six-inch (15-cm) telescope in dark skies. It's an example of an Sb-type spiral galaxy, according to Hubble's system of classification *(page 18)*.

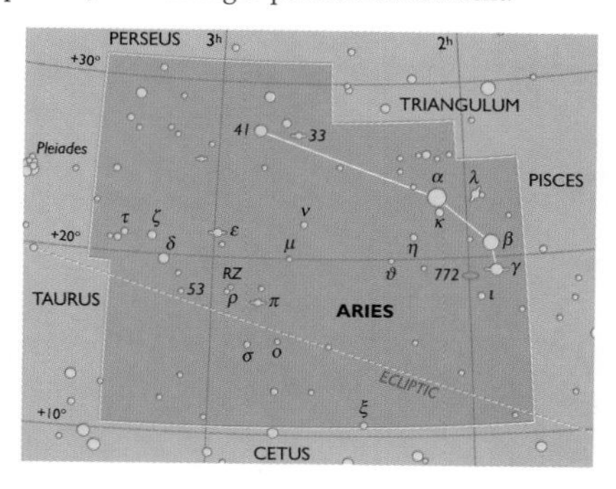

AURIGA (OH-RYE-GAH) | *The Charioteer*

• **Abbreviation:** Aur • **Genitive:** Aurigae • **Area:** 657 square degrees • **Size Ranking:** 21st

• **Best Viewed:** January to February • **Latitude:** 90ºN–34ºS • **Width:** • **Height:**

In Greek mythology, Auriga, the embodiment of Erechtheus, was a cripple who invented the horse-drawn chariot to travel about. When he died, the gods placed Erechtheus among the stars. This is one of the brightest constellations in the northern winter sky. All of its major stars are visible to the naked eye in an urban environment, and in a dark sky it is further star-spangled with many lesser-magnitude stars as well as the faint glow of the Milky Way, which covers most of the constellation. Objects of naked-eye interest include the first magnitude yellow giant star Capella, alpha (α) Aurigae, and immediately southeast of it, a triangular shape known as the Kids. Gamma (γ) Aurigae is shared with the constellation Taurus to the south, where the star is known as Beta (β) Tauri and delineates one of the Bull's horns. The whole region is a wonderful area to sweep with a pair of binoculars. Auriga also contains three fine Messier objects: M36, M37, and M38.

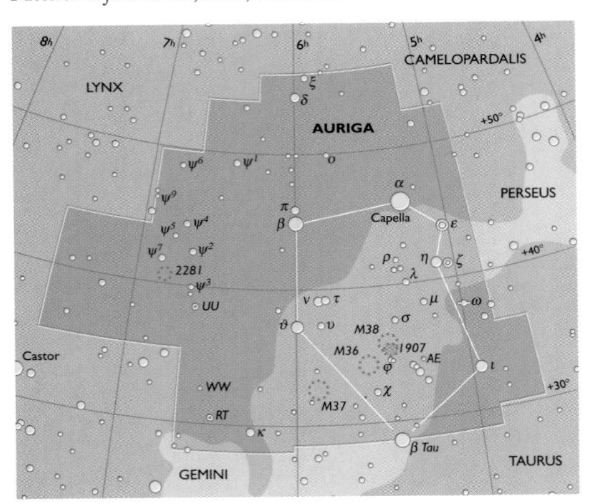

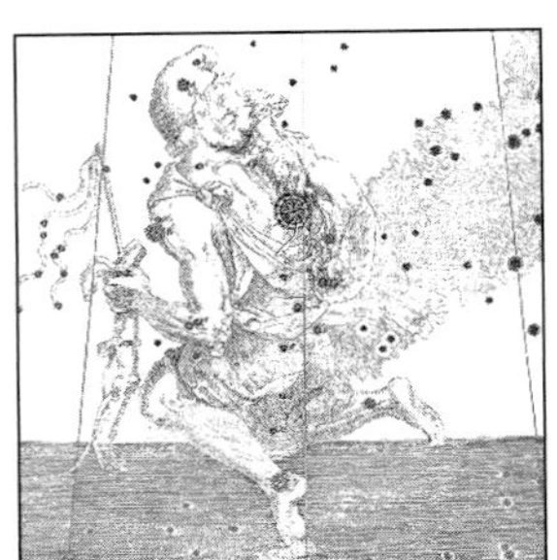

The stars of Auriga defined a chariot for ancient astronomers. The constellation was one of the first to be created.

FEATURES OF INTEREST

EPSILON (ε) AURIGAE: This is one of the most impressive variable stars visible in the night sky. The brightness varies between magnitudes +2.94 and +3.83 over a period of more than twenty-seven years. The dimming occurs during a twenty-three-month period, when a companion star passes in front of it, eclipsing the light of Epsilon Aurigae.

ZETA (ζ) AURIGAE: Another eclipsing variable, with a change of brightness that is much less that Epsilon Aurigae—only 0.1 magnitude. The eclipse lasts for forty days.

MESSIER 36, 37, 38: These three clusters are very similar in brightness (ranging from +5.6 to +6.4) and are easily seen with binoculars, appearing pretty much round and nebulous. With a small telescope, however, the clusters are an interesting study in contrasts. M36 is the coarsest of the three, consisting of about sixty stars (the brightest: magnitude +8.86), and also the smallest (diameter: twelve arc minutes). M37 is the richest of the three (at least 150 stars). Its members are all remarkably similar in brightness with a prominent, brighter red star at the center. The impression is one of glittering diamond dust on a velvet black background. M38 is similar to M36, though it is larger (diameter: about twenty arc minutes) and richer in stars (at least one hundred are visible in a six-inch (15-cm) telescope). M38 is notable for its cross-shaped structure of bright stars, which can be seen with a small telescope. Immediately south of Messier 38 is the smaller, fainter cluster NGC 1907, viewable with eight-inch [20-cm] and larger telescopes.

NGC 2281: A coarse cluster of about thirty stars—a good example of the many other minor open clusters visible in this constellation.

BOÖTES (BOH-OH-TEEZ) | *The Herdsman*

• **Abbreviation:** Boo • **Genitive:** Boötis • **Area:** 907 square degrees • **Size Ranking:** 13th

• **Best Viewed:** May to June • **Latitude:** 90ºN–35ºS • **Width:** 🖐✋ • **Height:** 🖐🖐✋

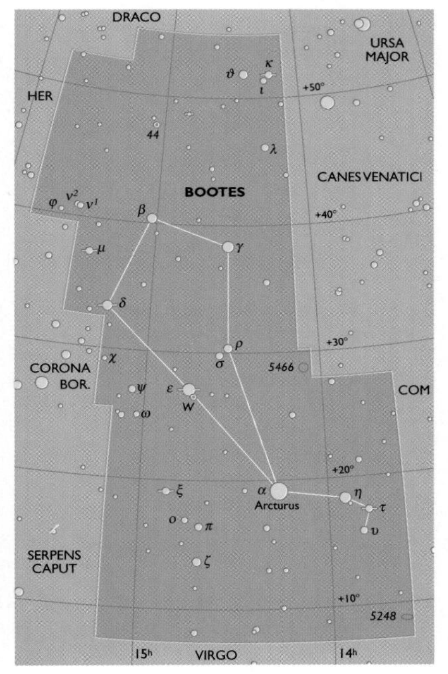

The most enduring legend surrounding this constellation is that Boötes is a hunter who, with his hound dogs, eternally pursues the Great Bear as it circles the pole. A giant kite-shaped group of stars, this prominent northern spring constellation is dominated by the brilliant orange star Arcturus, alpha (α) Boötes. At magnitude -0.1, it is the brightest star north of the celestial equator.

FEATURES OF INTEREST

EPSILON (ε) BOÖTIS: A colorful binary system that is difficult to resolve with a small telescope owing to the closeness of the components and the significant difference in brightness. The primary is magnitude +2.70; the secondary, magnitude +5.12. The separation is three arc seconds.

MU (μ) BOÖTIS: A triple star suitable for binoculars: magnitudes +4.3 and +6.5.

XI (ξ) BOÖTIS: An attractive telescopic binary system, magnitudes +4.7 and +6.8. The system has an orbital period believed to be about 150 years with separation of the components varying between 1.8 and 7.3 arc seconds. It was at its widest separation in 1984 and is now closing.

NGC 5248: This magnitude +10.3 spiral galaxy is the brightest one that Boötes has to offer and requires at least a six-inch (15-cm) aperture telescope to display its hazy elongated oval of light.

CAELUM (SEE-LUM) | *The Chisel*

• **Abbreviation:** Cae • **Genitive:** Caeli • **Area:** 125 square degrees • **Size Ranking:** 81st

• **Best Viewed:** December to January • **Latitude:** 41ºN–90ºS • **Width:** ✋ • **Height:** 🖐

This is one of the smallest, least conspicuous constellations in the sky, another of the groups first delineated by the French astronomer Lacaille in 1752. For the amateur skywatcher, the constellation contains a handful of faint galaxies and not much else.

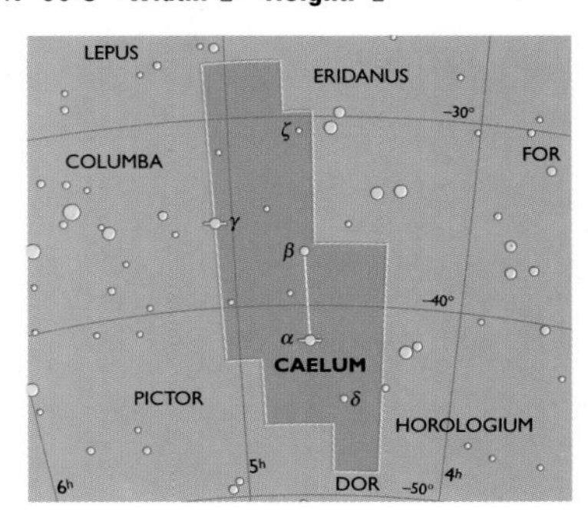

FEATURES OF INTEREST

ALPHA (α) CAELI: Although the components of this double star are separated by 6.6 arc seconds, this is a difficult object to split owing to the faintness of the secondary, magnitude +13. Not suitable for telescopes with less than a twelve-inch (30-cm) aperture.

GAMMA (γ) CAELI: A double star that can be seen with a three-inch (8-cm) telescope. The components are magnitude +4.5 and +8, separated by 2.9 arc seconds.

CAMELOPARDALIS (KA-MEL-O-PAR-DA-LIS) | *The Giraffe*

• **Abbreviation:** Cam • **Genitive:** Camelopardalis • **Area:** 757 square degrees • **Size Ranking:** 18th

• **Best Viewed:** December to May • **Latitude:** 90ºN–3ºS • **Width:** • **Height:**

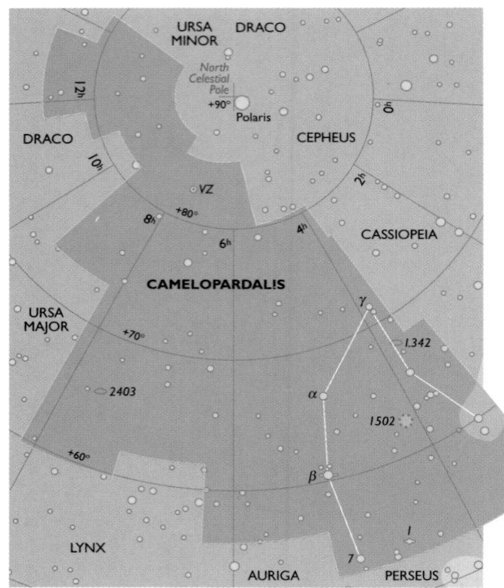

One of only a handful of constellations in the northern hemisphere not first recognized by the ancients, Camelopardalis has been called "the absence of a constellation." None of its stars is brighter than fourth magnitude.

FEATURES OF INTEREST

BETA (β) CAMELOPARDALIS: A four-magnitude yellow supergiant, Beta Camelopardalis is a double star with a 8.6-magnitude companion.

NGC 1502: A well-defined cluster that is a good target for binoculars and small telescopes.

NGC 2403: Binocular observers can spot this hazy oval of light in dark skies. Small telescopes (three-inch [8-cm])) reveal a bright galaxy; this is the fourteenth brightest celestial object of its type visible from Earth.

CANCER (CAN-SER) | *The Crab*

• **Abbreviation:** Cnc • **Genitive:** Cancri • **Area:** 506 square degrees • **Size Ranking:** 31st

• **Best Viewed:** February to March • **Latitude:** 90ºN–57ºS • **Width:** • **Height:**

This constellation has a rich mythological tradition. One legend relates that Cancer was the portal through which souls came to Earth and inhabited the bodies of men. In Greek mythology, it was the crab that clung to Hercules as he battled the water snake of the Lernaean marsh. Although none of its stars top the fourth magnitude, the constellation is easy to locate, lying between two bright neighbors—Gemini to the west and Leo to the east.

FEATURES OF INTEREST

IOTA (ι) CANCRI: A double star, magnitudes +4.5 and +6.5. Separated by 30.5 arc seconds, this attractive pair appears yellow and blue.

MESSIER 44: This star cluster is also known by the proper names Praesepe or the Beehive. Visible in dark skies to the naked eye as a misty patch of light, the cluster is well resolved in 7x50 binoculars. It is best seen in a small telescope at low magnification.

MESSIER 67: This is a fine open cluster, moderately bright and fairly compressed, well seen in a telescope of six- to eight-inches (15- to 20-cm) aperture. M67 is one of the oldest open clusters in the Milky Way; astronomers estimate its age at ten billion years.

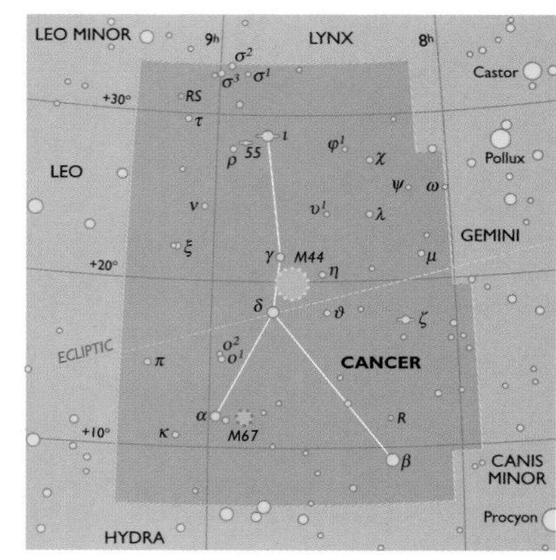

CANES VENATICI (KAY-NEEZ VE-NAT-EH-SEE) | *The Hunting Dogs*

• **Abbreviation:** CVn • **Genitive:** Canum Venaticorum • **Area:** 465 square degrees • **Size Ranking:** 38th

• **Best Viewed:** April to May • **Latitude:** 90ºN–37ºS • **Width:** • **Height:**

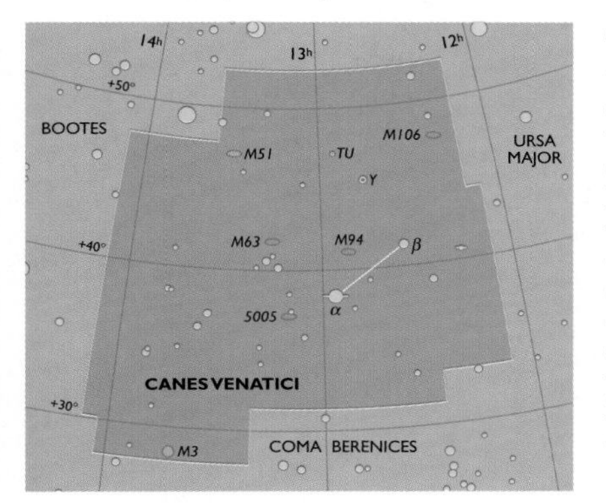

This constellation represents the hunting dogs that eternally accompany Boötes in his pursuit of Ursa Major, the Great Bear. Although only two of the stars in the constellation are bright enough to have Greek letter designations, the International Astronomical Union has been very generous in drawing the boundaries of the constellation so that it actually covers a large area of sky. This is a happy hunting ground for experienced amateur astronomers, as the constellation is home to dozens of galaxies easily visible with moderate-sized telescopes. Five Messier objects are located here: Messier 3, 51, 63, 94, and 106.

FEATURES OF INTEREST

ALPHA (α) CANUM VENATICORUM: Also known as Cor Caroli, this is an easy double star for even a small telescope, owing to its wide separation of 19.6 arc seconds.

MESSIER 3: Located halfway between Cor Caroli and Arcturus, this is one of the most spectacular globular clusters in the northern hemisphere, outranked only by M13 and M5. Although bright, the cluster does not resolve easily into individual stars with small telescopes. A six-inch (15-cm) telescope at high power is required to show stars around its outer edges.

MESSIER 51: The Whirlpool galaxy is the most spectacular example in the heavens of a spiral galaxy viewed face on. With the smaller irregular galaxy NGC 5195, it seems to form a faint double nebula in a small telescope, with the core of M51 appearing a little larger than its companion galaxy. The careful observer will notice traces of the Whirlpool's spiral structure with an eight-inch (20-cm) telescope.

MESSIER 63: Sometimes known as the Sunflower galaxy, this is another bright galaxy, appearing as an elongated haze with a very bright core as viewed by small telescopes. The outer spiral structure, which is heavily mottled, is only visible with large amateur telescopes. Magnitude: +8.6.

MESSIER 94: Although small, this magnitude +8.1 spiral galaxy has a core that is extremely bright. In a small telescope, M94 looks like a bright, nebulous star. Seen almost face on, like M63 *(above)*, it features very faint outer arms.

MESSIER 106: At magnitude +8.3, this bright object is also the visually largest of the Messier galaxies in Canes Venatici. It has a very large core and two strong, symmetrical spiral arms, the structure of which can be seen with large amateur telescopes. While in the vicinity, the observer can try to track down some of the fainter NGC galaxies surrounding M106; there are ten within one degree of this Messier object.

NGC 5005: This spiral galaxy, magnitude +9.8, is a fine specimen for small telescopes and a good representative of the many bright NGC galaxies in Canes Venatici.

M51
The Whirlpool galaxy is located in the northwest part of the constellation, near the star that lies at the end of the Big Dipper's handle.

CANIS MAJOR (KAY-NISS MAY-JER) | *The Greater Dog*

• **Abbreviation:** CMa • **Genitive:** Canis Majoris • **Area:** 183 square degrees • **Size Ranking:** 43th

• **Best Viewed:** January to February • **Latitude:** 56ºN–90ºS • **Width:** • **Height:**

Canis Major was recognized as a dog by all ancient middle-eastern civilizations and was seen, along with Canis Minor, as the faithful companion of Orion the Hunter. In Greek mythology, it was associated with the dog Laelaps, which raced a fox and won. This is a bright and easily recognized constellation featuring the sky's brightest star, Sirius the Dog Star. The Milky Way crosses the eastern region of the constellation and, as a result, Canis Major is well known for its numerous open clusters, many of which are suitable for binoculars or a small telescope. Canis Major is home to one Messier object, the open cluster M41.

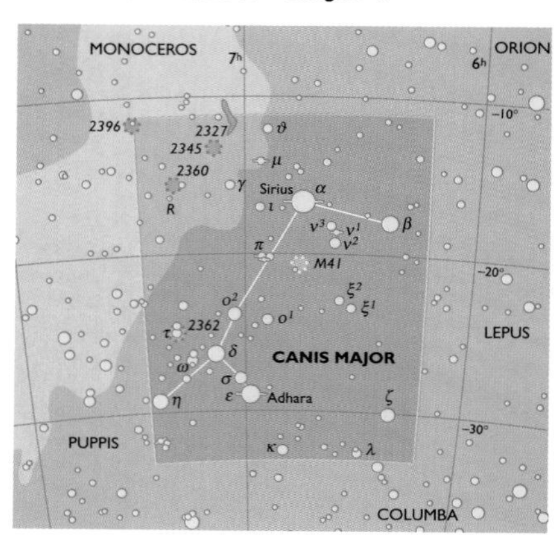

FEATURES OF INTEREST

ALPHA (α) CANIS MAJORIS: Better known as Sirius, this star is noteworthy for being the brightest star in the sky, though the distinction is more a result of its close proximity to our solar system. At a distance of 8.6 light-years, Sirius is the seventh closest star to the sun. It is also a binary star, though unfortunately not a suitable target for the small telescope owner. The secondary, a white dwarf star discovered by the nineteenth century telescope maker Alvan Clark, is comparatively faint and close to the primary. In a telescope, Sirius appears brilliant white. It flashes when low on the horizon, and its light is affected by disturbances in the Earth's atmosphere.

MESSIER 41: This large cluster is readily seen and partially resolved in binoculars, located about four degrees almost due south of Sirius. M41 is a fairly large cluster with ten stars brighter than magnitude +8.5. With a small telescope, use the lowest magnification available to get the best view. A prominent red star lies near the center of the field.

NGC 2327: Part of a large, faint emission nebula designated IC 2177, NGC 2327 is a slightly more conspicuous portion of this nebula complex—not a separate object. Beginners with small telescopes will have difficulty observing this object.

NGC 2345: This is an open cluster with about seventy stars, seven of which are about magnitude +9. The rest are fainter than magnitude +11.

NGC 2360: A spectacular cluster, one of the finest in Canis Major. As observed by an eight-inch (20-cm) telescope, the cluster is a small, well-resolved, very compressed grouping, elongated east/west, with a conspicuous spur angling off to the southeast. About fifty stars can be seen, seeming like glittering dust on a faint, background glow.

NGC 2362: This cluster is easily found as it surrounds Tau (τ) Canis Majoris, which is visible to the naked eye. It is a brilliant grouping of thirty to forty stars similar in brightness. NGC 2362 is less than a million years old, making it one of the youngest star clusters known. Tau Canis Majoris is strongly suspected of being one of its members. If so, it is one of the most brilliant stars in our galaxy, outshining our sun by a factor of fifty thousand.

The sky's brightest star, Sirius clearly outshines its stellar neighbors in Canis Major.

CANIS MINOR (KAY-NISS MY-NER) | *The Lesser Dog*

• **Abbreviation:** CMi • **Genitive:** Canis Minoris • **Area:** 380 square degrees • **Size Ranking:** 71st

• **Best Viewed:** February • **Latitude:** 89ºN–77ºS • **Width:** • **Height:**

Widely likened to a dog in ancient times, this is the lesser companion to Orion the Hunter. Canis Minor is a very small constellation. Only two stars are visible to the naked eye from urban locations. One is Procyon, the eighth brightest star in the sky. There is very little of interest to the small-telescope owner here.

FEATURES OF INTEREST

ALPHA (α) CANIS MINORIS: More commonly known as Procyon, this magnitude +0.4 star is 11.41 light-years from the sun. It is also a close binary star, though not suitable for a small telescope since the secondary, a white dwarf, orbits within five arc seconds of the primary and is a faint magnitude +10.7.

DOLIDZE 26: A loose cluster of stars located south and east from the sixth-magnitude star 6 Canis Minoris. With twelve-inch (30-cm) telescopes, about twenty stars, magnitude ranging from +10 to +12, are visible in an elongated patch from east to west.

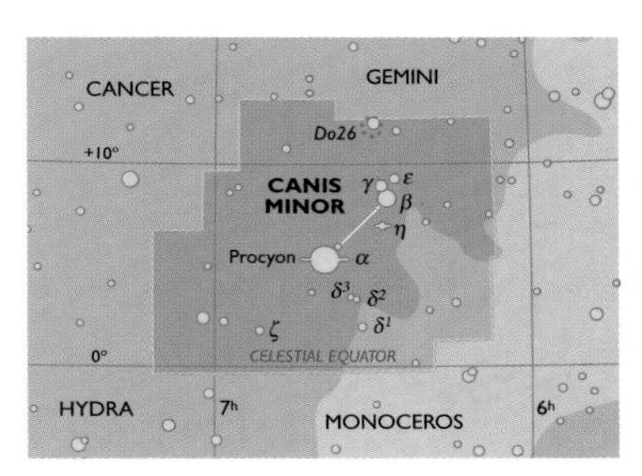

CAPRICORNUS (KAP-REH-KOR-NUSS) | *The Sea Goat*

• **Abbreviation:** Cap • **Genitive:** Capricorni • **Area:** 414 square degrees • **Size Ranking:** 40th

• **Best Viewed:** August to September • **Latitude:** 62ºN–90ºS • **Width:** • **Height:**

Capricornus the Sea Goat was thought to be the gateway of the souls of men as they ascended into heaven. This southern constellation is fairly large and conspicuous, though it contains no first-magnitude stars. Its outline is fairly distinctive, however, and can be easily traced from dark sky locations. Despite its great size, Capricornus contains only a few sights that will interest the amateur astronomer. The galaxies within its confines are all faint and inconspicuous. Capricornus does harbor one Messier object—the globular cluster M30.

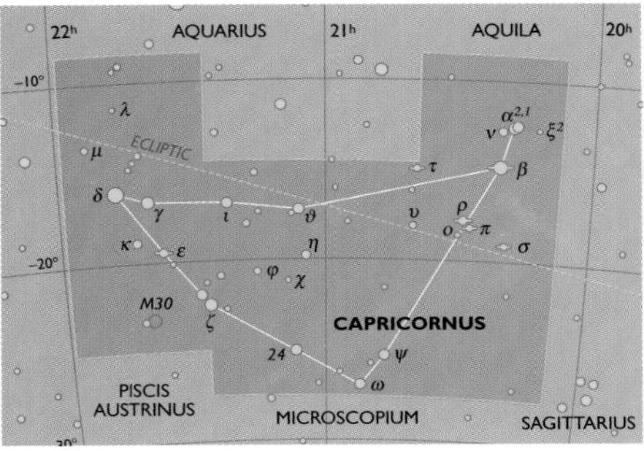

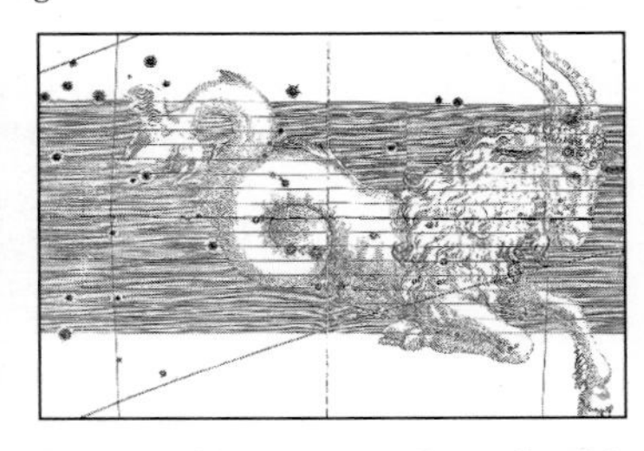

Depictions of Capricornus the Goat or Goat-Fish date back at least five thousand years.

FEATURES OF INTEREST

ALPHA (α) CAPRICORNI A naked-eye double star, better seen in binoculars. The stars are an optical double—not physically related but lying in the same direction, giving them the appearance that they are together. Each of these stars is, in turn, a double star when seen through a telescope. Alpha (α) 1 has two components, magnitudes +4 and +9, separated by forty-five arc seconds. Alpha (α) 2 is more difficult: magnitudes +3.5 and +11 with a separation of 6.6 arc seconds.

MESSIER 30: This magnitude +7.5 globular cluster seems less than six arc minutes across in a small telescope; to resolve the edges, a six-inch (15-cm) scope is needed. Distance: twenty-four thousand light-years. Careful observers will note two rows of stars emerging to the north of the cluster's center.

CARINA (KA-RYE-NAH) | *The Keel*

• **Abbreviation:** Car • **Genitive:** Carinae • **Area:** 494 square degrees • **Size Ranking:** 34th

• **Best Viewed:** January to April • **Latitude:** 14ºN–90ºS • **Width:** • **Height:**

In the late 1870s the astronomer Benjamin Gould at the Cordoba Observatory in Argentina embarked on a reformation of the ancient southern constellation Argo Navis. He divided it into four separate constellations: Puppis the Stern, Pyxis the Compass, Vela the Sails, and Carina the Keel. Carina is a magnificent constellation—a wonder either with or without optical aid. A bright section of the Milky Way crosses its eastern regions, and there are a wide variety of objects here of interest to the amateur astronomer, including several large, bright star clusters and the brightest nebula complex in the sky, NGC 3372—Eta (η) Carinae Nebula. The constellation is also home to the second brightest star in the sky, Canopus. Unfortunately for observers in mid-northern latitudes, this rich region of the sky is forever hidden below the southern horizon.

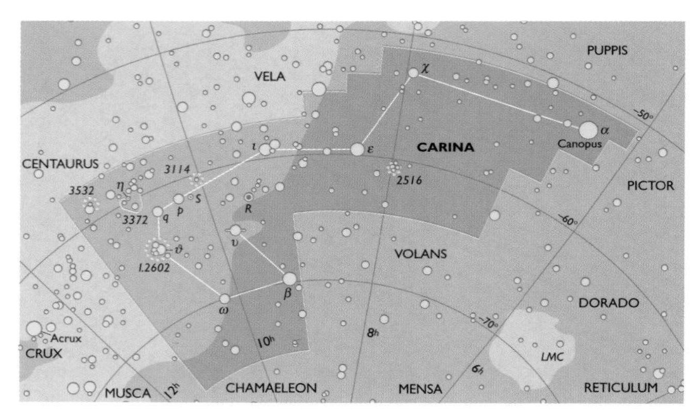

FEATURES OF INTEREST

ALPHA (α) CARINAE (CANOPUS): Unlike Sirius, which is bright because it is so close to our solar system, Canopus really *is* a brilliant star. With a magnitude of -0.72, it is fainter than some of our galaxy's beacons, such as Rigel and Deneb, but not by much.

UPSILON (υ) CARINAE: A double star, visible with a telescope. The magnitude of the primary is +3.01; the secondary, magnitude +6.26. The two are separated by five arc seconds.

R CARINAE: A variable star, ranging in magnitude from +4 to +10 over a period of 309 days.

S CARINAE: Another long-period variable with a magnitude that ranges from +5.4 to +9.5 over 149 days.

NGC 2516: Visible to the naked eye, this large open cluster is about fifty arc minutes in diameter and contains at least a hundred members, roughly twenty of them brighter than magnitude +9. Three of the stars are bright orange, and there are a few double- and triple-star systems involved.

NGC 3114: This cluster is a little smaller than NGC 2516, but it is just as noteworthy, featuring about a hundred stars with magnitudes ranging from +9 to +13. NGC 3114 supplies excellent viewing for binoculars and for a small telescope equipped with a low-power eyepiece.

NGC 3372: The Eta (η) Carinae Nebula is more than two degrees in diameter and, with its associated star clusters and bright nebulae, is arguably the most extraordinary region in the Milky Way. Also known as the Keyhole Nebula,

Lying in a rich part of the Milky Way, Carina offers a variety of clusters and nebulae suitable for binocular and telescope users. Its brightest star, Canopus, can be seen near the top right.

NGC 3372
Four times the apparent width of the full moon, NGC 3372 is best seen with binoculars.

it is easily spotted by the naked eye. Photographs show at least four major portions, divided by dark channels. The nebula is home to the extraordinary variable star Eta (η) Carinae, a likely candidate to become a supernova in the near future. This star is currently not visible to the naked eye, but over a period of 270 years beginning in 1677, the star underwent a number of extraordinary brightness variations. In 1843 it briefly became the second brightest star in the sky at magnitude –0.8. The star is part of a tiny nebula known as the Homunculus, which has been photographed in extraordinary detail by the Hubble Space Telescope.

NGC 3532: Easily visible to the naked eye, this is probably the finest open cluster in the heavens—certainly the best in the southern hemisphere. This cluster measures about sixty by thirty arc minutes in size and contains at least 150 stars brighter than magnitude +12.

IC 2602: At least thirty stars are visible in a seventy-five-arc-minute diameter surrounding the third magnitude star Theta (ϑ) Carinae. This is an excellent binocular object.

IC 2602
This bright open cluster surrounding Theta Carinae is also known as the Southern Pleiades.

CASSIOPEIA (KAS-EE-OH-PEE-UH) | *The Queen*

• **Abbreviation:** Cas • **Genitive:** Cassiopeiae • **Area:** 598 square degrees • **Size Ranking:** 25th

• **Best Viewed:** October to December • **Latitude:** 90ºN–12ºS • **Width:** • **Height:**

In Greek mythology, Cassiopeia was the wife of King Cepheus of Ethiopia. A vain and boastful woman, she was punished for her hubris by the god of the sea, Poseidon, who sent a sea monster to ravage the coast of Ethiopia. This bright north constellation is one of the sky's most recognizable groupings of stars: The five brightest stars form the outline of the letter W. Owing to its proximity to the Milky Way, this is a beautiful area of the sky to explore with a pair of binoculars. It contains many open clusters as well as several well-known but faint nebulae. Cassiopeia holds two Messier objects: the open clusters M52 and M103.

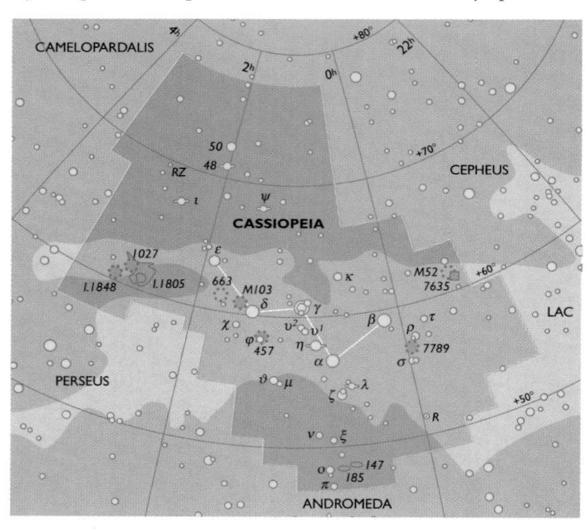

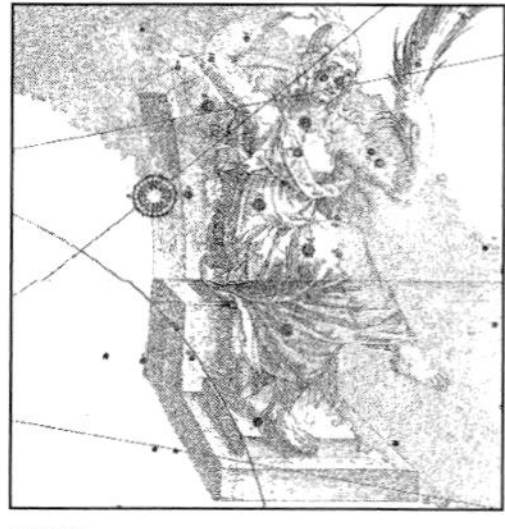

M52

Located at the western edge of the constellation, M52 is an easily seen example of an open cluster. It lies fifty-two hundred light-years away.

FEATURES OF INTEREST

GAMMA (γ) CASSIOPEIAE: This is an erratic variable star that has undergone slow but definite brightness changes over the last ninety years. It brightened slowly between 1910 and 1937, reaching a magnitude of +1.6 magnitude before fading to third magnitude three years later. It then brightened to about magnitude +2.2 by the mid-1970s. Currently, it is about magnitude +2.5. Gamma Cassiopeiae, which lies at the center of the W, is also a double star, but you'll need at least an eight-inch (20-cm) telescope to resolve the two. The secondary is magnitude +8.8, separated by only two arc seconds.

ETA (η) CASSIOPEIAE: An excellent binary star for the small telescope user. Component A: magnitude +3.44; component B: magnitude +7.51. The gap between them—twelve arc seconds—is slowly widening.

IOTA (ι) CASSIOPEIAE: A beautiful, tight triple star system. Component A is magnitude +4, component B, magnitude +7. The two are separated by 2.2 arc seconds. Component C is magnitude +8, separated by 7.3 arc seconds.

RHO (ρ) CASSIOPEIAE: A bright yellow supergiant that pulsates, oscillating from magnitude 4 and 5 roughly every year.

MESSIER 52: This is a rich, bright, and fairly compressed cluster of stars with a total apparent magnitude of +6.9 and a diameter of about twelve arc minutes. Large amateur telescopes will reveal about a hundred stars down to magnitude +13. The intriguing Bubble Nebula, NGC 7635, is located about thirty-five arc minutes to the southwest. As its name suggests, the Bubble Nebula looks as though it harbors an inflated ball at its center.

MESSIER 103: A magnitude +7.4 cluster with stars of varying brightnesses. About thirty to forty stars are visible, ranging from magnitudes +8 to +12.

NGC 147/185: This pair of dwarf elliptical galaxies are distant satellites of the Andromeda galaxy. NGC 147 is larger though fainter than NGC 185. It features a small, bright core. NGC 185 appears as a diffuse glow that brightens smoothly to its core.

NGC 457: This bright and attractive cluster is easily located, since it surrounds the naked-eye star Phi (φ)Cassiopeiae. The cluster is also called the Night Owl because in small telescopes the outline really does look like a bird, with Phi and a fainter cluster member representing the eyes and two spurs of stars representing the wings. At least thirty stars to magnitude +12 are visible, with a compact triangle of three stars near the center.

NGC 663: A fairly rich and compact cluster, elongated north/south. Most of the stars hover around +10 to +11 magnitude.

NGC 1027: A small, weak cluster surrounding a magnitude +8 field star. About twelve members are visible.

IC 1805/IC 1848: Located together along an east/west axis in the sky, these are large, faint nebula requiring an ultra-high-contrast (UHC) filter and a low-power instrument to see. IC 1805 surrounds the scattered cluster Melotte 15.

NGC 7789: This is an extraordinarily rich open cluster, easily seen in binoculars as a faint nebulous patch a little over half the size of the full moon. More than two hundred cluster members can be detected in eight- to twelve-inch (20- to 30-cm) telescopes in dark skies. The cluster members are for the most part quite similar in brightness.

M52
Located at the western edge of the constellation, M52 is an easily seen example of an open cluster. It lies fifty-two hundred light-years away.

NGC 7635
This faint planetary nebula, called the Bubble Nebula, is only visible to twelve-inch (30-cm) and larger telescopes.

CENTAURUS (SEN-TOR-US) | *The Centaur*

• **Abbreviation:** Cen • **Genitive:** Centauri • **Area:** 1,060 square degrees • **Size Ranking:** 9th

• **Best Viewed:** April to June • **Latitude:** 25ºN–90ºS • **Width:** • **Height:**

With its head and torso of a man and the hind quarters of a horse, the Centaur is a well-known mythological figure. This constellation is one of two that depict the creature; the other is Sagittarius the Archer. To the ancient Greeks, Centaurus was identified with Chiron, a pacific creature, opposite in character to the aggressive Sagittarius, who is always depicted with bow and arrow drawn. A large and bright constellation, Centaurus has two first-magnitude stars, Alpha (α) and Beta (β) Centauri. This region of the sky has a rich body of astronomical objects of interest for users of both modest and sophisticated optical equipment.

NGC 5139
In 1677 Edmond Halley was the first astronomer to realize that the fuzzy patch in Centaurus was really a cluster of stars.

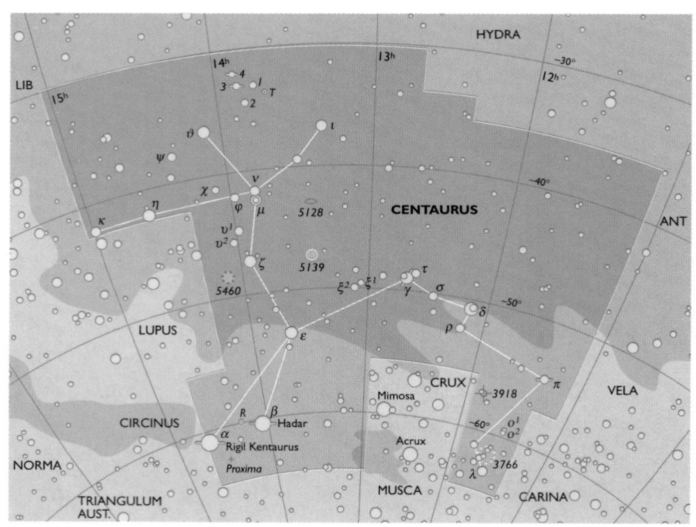

FEATURES OF INTEREST

ALPHA (α) CENTAURI (RIGEL KENTAURUS): This is a triple-star system with the distinction of being the sun's nearest stellar neighbors. The primary star, Alpha Centauri A, is identical in chemical composition and brightness to the sun. Alpha Centauri (magnitude 0) has a companion, magnitude +1.33, located 13.8 arc seconds away. The two orbit each other every eighty years. The third member, Proxima Centauri, is two degrees distant to the southwest. This is the closest star to Earth—4.2 light-years away, a red dwarf of magnitude 12.4.

BETA (β) CENTAURI: A blue-white giant, magnitude 0.6.

GAMMA (γ) CENTAURI: Twin third-magnitude stars, which are very difficult to separate with a small telescope as the pair are never more than 1.7 arc seconds apart.

NGC 3766: Visible to the naked eye as a misty patch of light, this is an excellent open cluster for the small telescope. Magnitude: +4.4. About sixty stars are visible in a twelve-arc-minute area. NGC 3766 is a very young cluster of stars, probably only ten million years old. It is located about fifty-eight hundred light years from Earth.

NGC 3918: This is a very bright, though small, planetary nebula with a strong bluish cast to its light. Diameter: twelve arc seconds; magnitude: +8.1.

NGC 5128: This bright galaxy lies very close to the Local Group, a collection of two dozen galaxies that includes our home galaxy, the Milky Way. It is is easily seen with even the smallest telescopes. NGC 5128 is a strong source of radio waves. It is also one of the most peculiar looking galaxies in the sky: a large elliptical system bisected by a prominent dust lane. The band is darker toward the southeast and can be seen with apertures as small as three inches (8 cm).

NGC 5139: This extraordinary globular cluster was known even to the ancients, who saw it as a fuzzy star. By the seventeenth century it had earned a Greek alphabet designation: Omega (ω) Centauri. It is unchallenged in the sky as the largest, brightest example of its class. Even the smallest telescope will begin to resolve individual stars, while larger apertures will reveal a breathtaking collection of thousands of stars visible across an unresolved disc of distant stars.

NGC 5460: A bright, scattered open cluster, with about twenty-five stars of magnitude +8 and fainter.

CEPHEUS (SEE-FEE-US) | *The King*

• **Abbreviation:** Cep • **Genitive:** Cephei • **Area:** 588 square degrees • **Size Ranking:** 27th

• **Best Viewed:** September to October • **Latitude:** 90ºN–1ºS • **Width:** • **Height:**

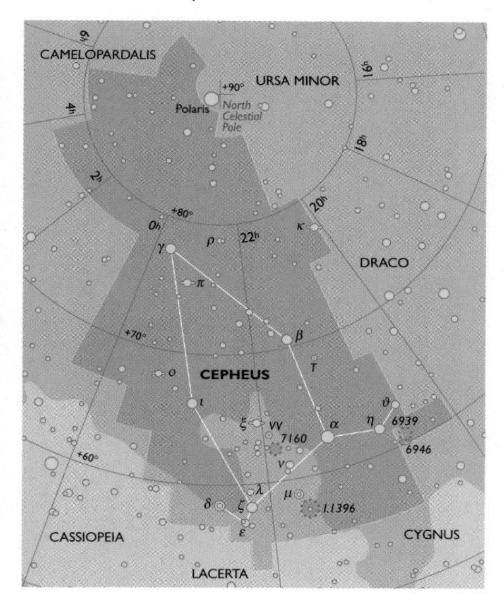

Cepheus is one of the oldest constellations in the heavens, a well-known sight for the Egyptians, the Arabs, and the ancient civilization that arose around the Tigris and Euphrates rivers around 2000 B.C. In legend, Cepheus was the king of the Ethiopians and Cassiopeia was his queen. He is also the father of Andromeda, chained to a rock in the sea monster's path as part of the punishment for her mother's vanity. Though not particularly bright, the outline of the constellation is easily recognized, resembling a child's simplistic drawing of a house.

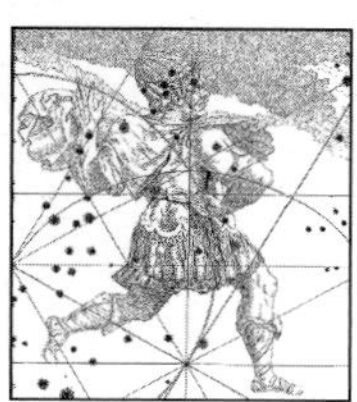

From the time of the ancient astronomers, the stars of Cepheus have been identified with the king of the Ethiopians.

FEATURES OF INTEREST

DELTA (δ) CEPHEI: This is the prototype star of a peculiar class of variables known as the Cepheids. These pulsating stars are used to estimate the distances of other celestial objects. Delta Cephei varies from magnitude +3.6 to +4.3 roughly every five days.

BETA (β) CEPHEI: A double and variable star with an eighth-magnitude companion.

MU (μ) CEPHEI: A red giant star, Mu Cephei's light varies over months or even years, ranging from about magnitude +3.7 to +5.0. Its reddish hue earned it the name of the Garnet Star.

NGC 6939: A delicate but attractive and rich star cluster, suitable for telescopes in the eight-inch (20-cm) range. About eighty stars are visible with magnitudes +12 and fainter. A low-power eyepiece will reveal the faint face-on spiral galaxy NGC 6946 about forty arc minutes southeast.

NGC 7160: A small coarse group of about a dozen stars, magnitudes +7 and fainter, about seven arc minutes in diameter.

CETUS (SEE-TUS) | *The Whale*

• **Abbreviation:** Cet • **Genitive:** Ceti • **Area:** 1,231 square degrees • **Size Ranking:** 4th
• **Best Viewed:** October to December • **Latitude:** 78ºN–78ºS • **Width:** • **Height:**

In Greek mythology, Cetus the Whale was the sea monster sent by the gods to ravage the coast of Ethiopia as a punishment to Queen Cassiopeia. The hero Perseus, fresh from slaying the only mortal Gorgon sister, Medusa, turned the sea monster to stone by showing him the Gorgon's hideous head. Cetus is a large constellation and the home of hundreds of galaxies, with many brighter examples for the small telescope. One Messier object is located here: the galaxy M77.

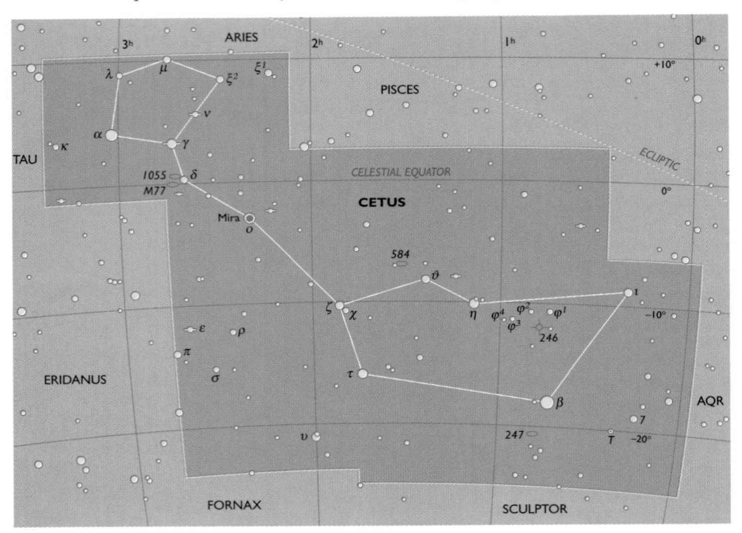

FEATURES OF INTEREST

OMICRON (ο) CETI (MIRA): Mira (Latin for "the wonderful") was the first of the long-period, pulsating variables to be discovered, by David Fabricius in the year 1596. Such stars—called Mira-class—feature a wide brightness range over a long period of time. In Mira's case, the magnitude varies from approximately 2 to 9 every 332 days.

GAMMA (γ) CETI: A close double star, magnitudes +3.5 and +6, separated by merely 2.7 arc seconds.

TAU (τ) CETI: The seventh closest star to the sun, 11.9 light-years distant, with a magnitude of +3.49.

MESSIER 77: This galaxy is the most distant object in Messier's catalog: eighty-one million light-years from Earth. It is also the brightest member of a class of galaxy known as Seyfert, named after the astronomer Carl Seyfert, who first noted spiral galaxies with uncommonly bright nuclei. M77 can be seen with 2.4-inch (6-cm) and larger instruments.

M77
In a small telescope, M77 appears as a faint disk surrounding a bright core.

NGC 246: Although the total magnitude of this planetary nebula is rather bright—about +8—it is nevertheless pale. Its diameter is almost four arc minutes, making it one of the larger examples of this class of object.

NGC 247: A large, faint galaxy that is part of a loose aggregation of galaxies that may form the nearest collection to our own Local Group. Distance: about ten million light-years.

CHAMAELEON (KA-MEE-LEE-UN) | *The Chameleon*

• **Abbreviation:** Cha • **Genitive:** Chamaeleontis • **Area:** 132 square degrees • **Size Ranking:** 79th

• **Best Viewed:** May to June • **Latitude:** 7ºN–90ºS • **Width:** • **Height:**

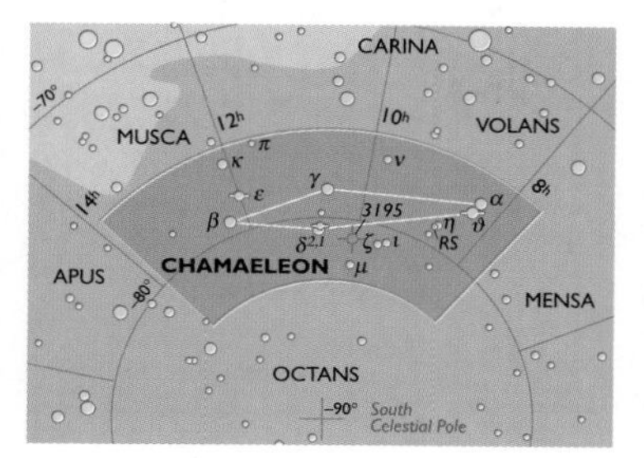

A dim constellation of the deep-southern skies, located between Carina and Octans, Chamaeleon was created by the stellar cartographer Johann Bayer in 1604.

FEATURES OF INTEREST

DELTA (δ) CHAMAELEONTIS: This is a wide, faint naked-eye pair, better seen in binoculars. The brighter of the two stars, Delta (δ) 1, is itself a very close double star.

NGC 3195: A planetary nebula that appears as a round, bluish disk. The central star, magnitude +15.3, is not visible with small telescopes.

CIRCINUS (SUR-SEH-NUS) | *The Drawing Compass*

• **Abbreviation:** Cir • **Genitive:** Circini • **Area:** 93 square degrees • **Size Ranking:** 85th

• **Best Viewed:** May to June • **Latitude:** 19ºN–90ºS • **Width:** • **Height:**

One of the French astronomer Nicolas de Lacaille's creations in the year 1752, this constellation features three principal stars of magnitudes +3 to +5. Circinus represents a dividing compass used by surveyors.

FEATURES OF INTEREST

ALPHA (α) CIRCINI: This is an attractive double star for the small telescope. The brilliant yellow primary is paired with a magnitude +9 red star located 15.7 arc seconds away.

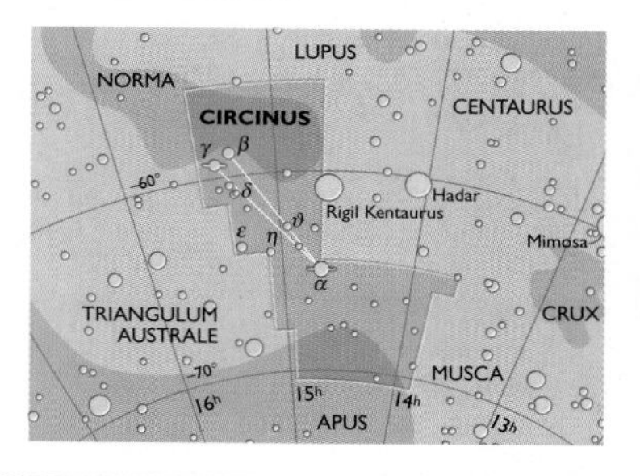

COLUMBA (KOH-LUM-BAH) | *The Dove*

• **Abbreviation:** Col • **Genitive:** Columbae • **Area:** 270 square degrees • **Size Ranking:** 54th

• **Best Viewed:** April to May • **Latitude:** 46ºN–90ºS • **Width:** • **Height:**

This small constellation represents the dove and was named by astronomer Petrius Plancius. Despite its location in the southern sky, the entire constellation is visible from mid-northern locations in the early evening in January and February.

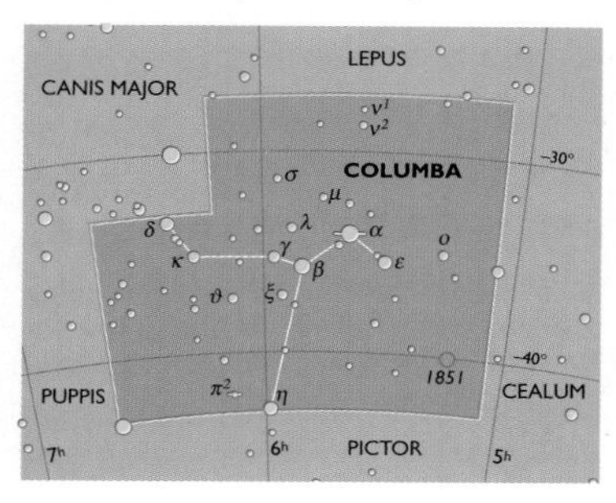

FEATURES OF INTEREST

MU (μ) COLUMBAE: This is one of the notorious threesome of Runaway Stars, which are traveling through the Milky Way at roughly sixty miles (100 km) a second. These stars appear to have been ejected by a violent explosion—possibly a supernova—from the Orion Nebula several million years ago. The other two members are 53 Arietis and AE Aurigae, the star that appears to illuminate the diffuse nebula IC 405.

NGC 1851: This is a very compressed globular cluster with a bright core.

COMA BERENICES (KOH-MAH BEAR-EH-NEE-SEEZ) | *Berenice's Hair*

• **Abbreviation:** Com • **Genitive:** ComBerenices • **Area:** 386 square degrees • **Size Ranking:** 42nd

• **Best Viewed:** April to May • **Latitude:** 90ºN–55ºS • **Width:** ✋ • **Height:** ✋

This constellation was well known to the ancients and has a wonderful story attached to its creation. According to legend, when King of Egypt Ptolemy Euergetes went off to war, his devoted young wife, Berenice, swore to the gods that she would sacrifice her beautiful tresses to Venus if her husband returned safely from the fighting. When he did, she cut off her hair, placing her locks in the temple. By the following day, they had disappeared. The temple guards were at the point of being put to death when the astronomer Conon approached the king and pointed to the glittering mass of stars in the heavens, saying that his wife's hair had been transferred to the stars. Coma Berenices is one of the richest areas in the sky for galaxy hunters and is the home of eight Messier objects, all but one of which are galaxies.

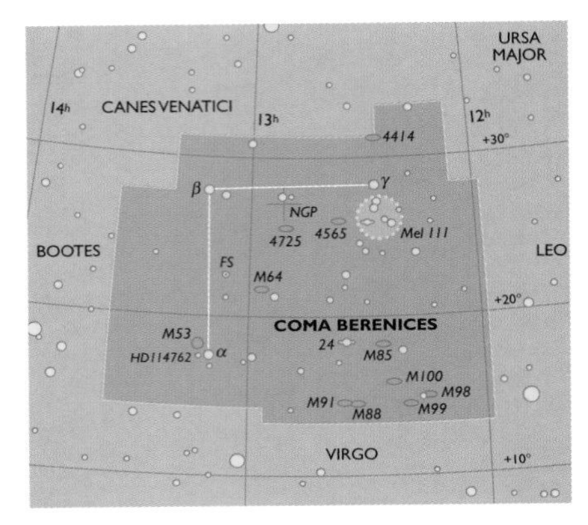

FEATURES OF INTEREST

MELOTTE 111: This is the Coma Berenices star cluster, one of the closest such objects to our solar system and easily visible to the naked eye. A beautiful sight with binoculars, it is too large to be seen well with a small telescope.

MESSIER 53: A bright globular cluster, very compressed, with partial resolution of the edges possible with an eight-inch (20-cm) telescope. While in the area, the observer can search forty-five arc minutes to the southeast for NGC 5053, a much fainter specimen, dimly visible with an eight-inch (20-cm) telescope.

MESSIER 64: A bright, smooth-armed spiral galaxy with a large black absorption cloud bordering the core to the north. This cloud of dust is detectable with six-inch (15-cm) and larger telescopes and gives M64 its nickname, the Blackeye galaxy.

MESSIER 88: A bright, highly tilted spiral galaxy visible with a three-inch (8-cm) telescope, notable for being the northeastern starting point of a remarkable archipelago of a cluster of galaxies called Markarian's Chain. Other Messier objects in the region include M85, M91, M98 (located near the star 6 Comae Berenices), M99, and M100.

NGC 4565: The finest example in the heavens of a spiral galaxy that can be seen edge on, and well worth seeking out although it is a rather dim specimen in a small telescope. A dark absorption cloud neatly bisects the galaxy along its rotational plane, which can be seen with eight-inch (20-cm) instruments as it crosses in front of the galaxy's core.

NGC 4565
This spiral galaxy, twenty million light-years from Earth, is similar in size to the Milky Way.

CORONA AUSTRALIS (KOR-OH-NAH OS-TRAY-LIS) | *The Southern Crown*

• **Abbreviation:** CrA • **Genitive:** Coronae Australis • **Area:** 128 square degrees • **Size Ranking:** 80th
• **Best Viewed:** July to August • **Latitude:** 44ºN–90ºS • **Width:** • **Height:**

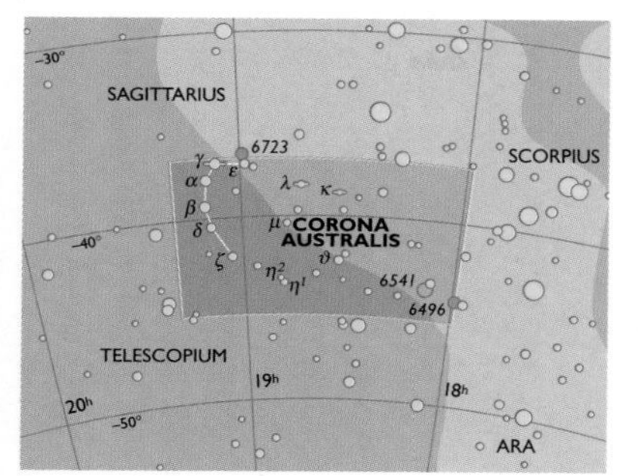

This southern constellation is a worthy complement to its northern cousin, Corona Borealis, although the Southern Crown is not quite as conspicuous, consisting primarily of stars magnitudes +4 to +6. About a dozen members form an obvious semicircle. Although the western region of the constellation is immersed in the Milky Way, there are no obvious deep-sky objects here. Still, the area is an interesting sweep for binoculars.

FEATURES OF INTEREST

GAMMA (γ) CORONAE AUSTRALIS: A binary star with two members that appear equal in brightness at magnitude +5. The orbital period is believed to be 122 years, and currently the pair is separated by only 1.2 arc seconds, making it difficult to resolve with telescopes smaller than six-inch (15-cm) aperture.

KAPPA (κ) CORONAE AUSTRALIS: This is a much easier double star, separation: 21.2 arc seconds, magnitudes: +6 and +6.5.

NGC 6541: A bright (magnitude +6.1) globular cluster with a fairly compressed core and a loosely structured outer halo. With the brightest members at magnitude +12.3, resolution is possible with six-inch (15-cm) and larger telescopes.

CORONA BOREALIS (KOR-OH-NAH BOR-EE-AL-ISS) | *The Northern Crown*

• **Abbreviation:** CrB • **Genitive:** Coronae Borealis • **Area:** 179 square degrees • **Size Ranking:** 73rd
• **Best Viewed:** June • **Latitude:** 44ºN–90ºS • **Width:** • **Height:**

This conspicuous star grouping was well known in antiquity and many ancient civilizations saw it as a crown or a wreath. The Greeks viewed it as the crown that Bacchus gave to Ariadne after she had been abandoned by Theseus. Seven stars, magnitudes +2 to +4 are involved in its outline. Although many galaxies populate the region, they are all beyond the reach of a small telescope.

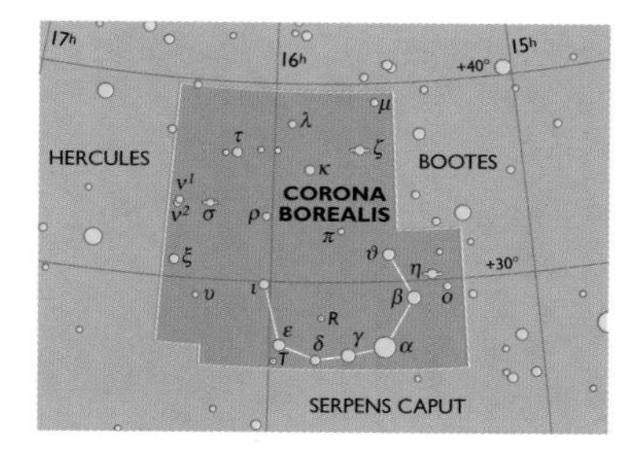

FEATURES OF INTEREST

RHO (ρ) CORONAE BOREALIS: A variable star known for peculiar and sometimes abrupt variations in brightness. Rho Cor Bor, as it is often called, stays near maximum light for many years and then suddenly fades in brightness to magnitude +15. The star is a supergiant with a strong presence of carbon in its spectrum, which suggests that dark clouds of carbon encircle the star, periodically blocking its light .

TAU (τ) CORONAE BOREALIS: Another remarkable variable star, this is one of the recurrent nova-class stars, which stays at minimum brightness for many years—often decades—before blazing forth. The mechanism at work is a transfer of mass between closely orbiting stars: a red giant and a smaller, massive blue star. The gravitational pull of the blue star steals material from the giant. After many years, the blue star's accumulation of mass reaches a critical point and the outer layer of the blue star literally explodes off the surface—after which the entire process is repeated again.

CORVUS (KOR-vus) | *The Crow*

• **Abbreviation:** Crv • **Genitive:** Corvi • **Area:** 184 square degrees • **Size Ranking:** 70th

• **Best Viewed:** April to May • **Latitude:** 65ºN–90ºS • **Width:** ✋ • **Height:** ✋

Most ancient civilizations depicted these stars as a bird. Greek mythology relates that Apollo fell in love with Coronis but was suspicious of her and sent a crow to spy on her. The crow told Apollo of her infidelity and for this service the bird was placed among the stars. The constellation is home to many galaxies, though most are not bright.

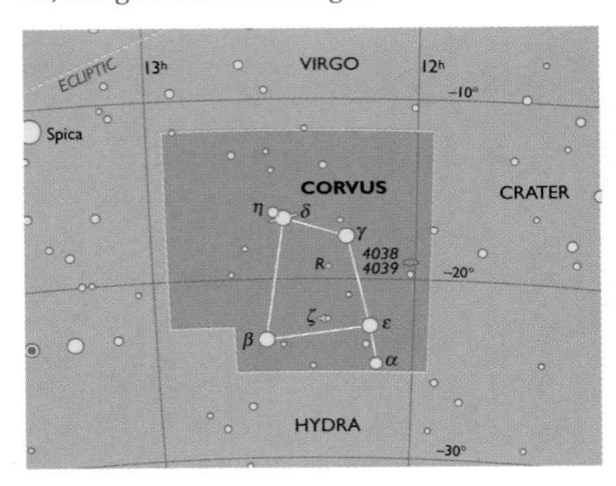

NGC 4038/4039
The Ring-Tailed Galaxy is actually two galaxies that are interacting, connected by the force of gravity.

FEATURES OF INTEREST

DELTA (δ) CORVI: A bright double star, suitable for a small telescope. Component A is magnitude +2.95; component B is a dwarf star, magnitude +8.26. Current separation of the components is twenty-four arc seconds.

NGC 4038/4039: This is the notorious Ring-Tailed Galaxy, one of the finest examples of two galaxies in collision, although little evidence of the extraordinary processes taking place are visible with a small telescope. The northern component is the brighter of the two and small telescopes will show it as a peculiar double nebula, joined together in the east. Astronomers have long been aware of the so-called Antennae, thin sprays of matter being ejected as a result of the collision. Closeup views obtained by the Hubble Space Telescope show dark clouds of dust and gas signifying intense star formation.

CRATER (KRAY-ter) | *The Cup*

• **Abbreviation:** Crt • **Genitive:** Crateris • **Area:** 282 square degrees • **Size Ranking:** 53rd

• **Best Viewed:** April • **Latitude:** 65ºN–90ºS • **Width:** ✋ • **Height:** ✋

One of Ptolemy's original forty-eight constellations, Crater is depicted as a cup that rests on the back of the sea serpent Hydra. According to Greek myth, Apollo asked Corvus to bring him a cup for a drink. Corvus did so, but only after stopping along the way to eat figs. He blamed his tardiness on Hydra. Apollo wasn't fooled by the excuse, however, and he placed Corvus to the west of Crater, just out of reach of the water in the cup. The area is of interest to galaxy hunters using ten-inch (25-cm) and larger telescopes.

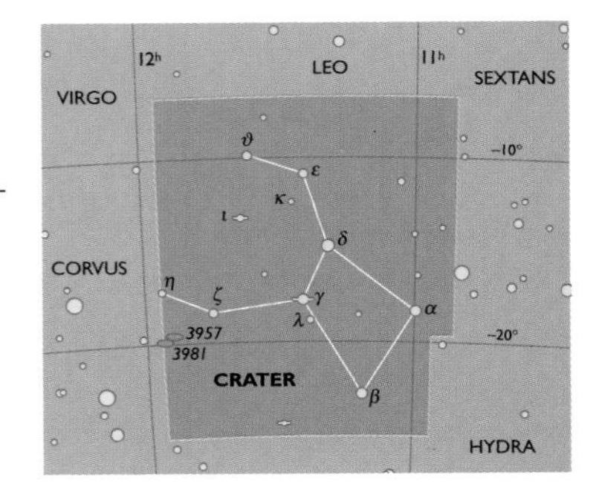

FEATURES OF INTEREST

GAMMA (γ) CRATERIS: A double star visible with three-inch (8-cm) and larger telescopes.

NGC 3957 AND NGC 3981: Two galaxies typical of the region. NGC 3957 is magnitude +12.6, a well-defined glow appearing edge on and oriented north-northwest/south-southeast. NGC 3981 is slightly brighter, but with diffuse, less well-defined edges.

CRUX (KRUCKS) | *The Southern Cross*

• **Abbreviation:** Cru • **Genitive:** Crucis • **Area:** 68 square degrees • **Size Ranking:** 88th

• **Best Viewed:** April to May • **Latitude:** 25ºN–90ºS • **Width:** • **Height:**

Despite its small size—it is the smallest of the constellations—Crux is an easily recognized star grouping located in an extraordinary region of the sky. The early Portuguese navigators who explored the southern oceans saw the cross-shaped star grouping as a symbol of their faith. The upright part of the cross points to the south celestial pole, which served once as a handy navigational guide for sailors. Crux is home to a brilliant star cluster and a famous dark nebula, and is a wonderful region to investigate with binoculars. Unfortunately, the constellation is hidden from the view of observers at mid-northern latitudes.

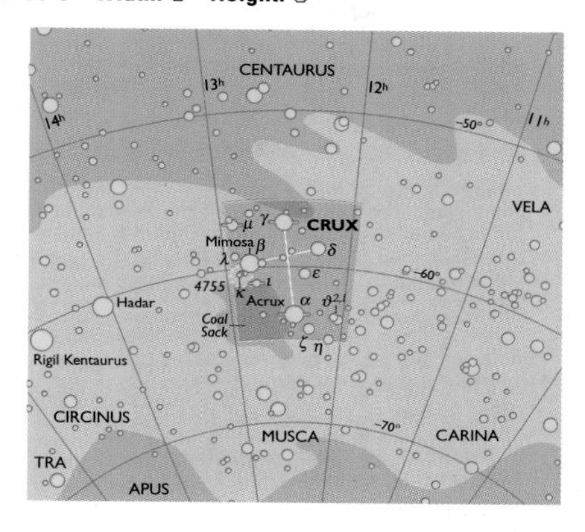

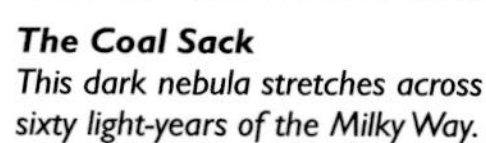

The Coal Sack
This dark nebula stretches across sixty light-years of the Milky Way.

FEATURES OF INTEREST

ALPHA (α) CRUXIS (ACRUX): A brilliant double, with stars of magnitudes +1.5 and +2, separated by four arc seconds. Small telescopes will require high magnification to split the two cleanly.

BETA (β) CRUXIS: A blue-white giant that pulsates in size every five hours, each time temporarily increasing its magnitude. There is an eleventh-magnitude companion star forty-four arc seconds away.

GAMMA (γ) CRUXIS: A bright orange magnitude 1.67 star that is 111 arc seconds from a magnitude 6.4 star. The two are not related.

NGC 4755: This is the Jewel Box, one of the more famous open clusters in the heavens, a brilliant assemblage dominated by very luminous stars that range in age from twenty-five to thirty million years. At least fifty stars are visible in an area less than twelve arc minutes in diameter, making it a suitable object for a small telescope.

THE COAL SACK: This is the most famous example of an absorption nebula, easily visible to the naked eye as a black stain set against the backdrop of the bright star clouds of the Crux Milky Way. Absorption nebulae are common in spiral galaxies and represent material from which stars will someday form. These nebulae consist of dust and gas that is not illuminated by the radiation of nearby stars.

CYGNUS (SIG-NUS) | *The Swan*

• **Abbreviation:** Cyg • **Genitive:** Cygni • **Area:** 804 square degrees• **Size Ranking:** 16th

• **Best Viewed:** August to September • **Latitude:** 90ºN–28ºS • **Width:** ✋✋ • **Height:** ✋✋

One of Ptolemy's original forty-eight constellations, this group of stars, known unofficially as the Northern Cross, has long been associated with birds. The ancient Arabs knew it as the Flying Eagle. In Greek mythology, Zeus took the form of a swan in order to seduce Leda, the wife of the king of Sparta. Another myth associates the group with Orpheus, transferred to the stars along with his lyre, an ancient musical instrument. For northern observers, Cygnus rests among one of the richest regions of the Milky Way. The slightest optical aid reveals countless resolved stars set against a softly luminous backdrop. Two Messier objects lie here: the open clusters M29 and M39.

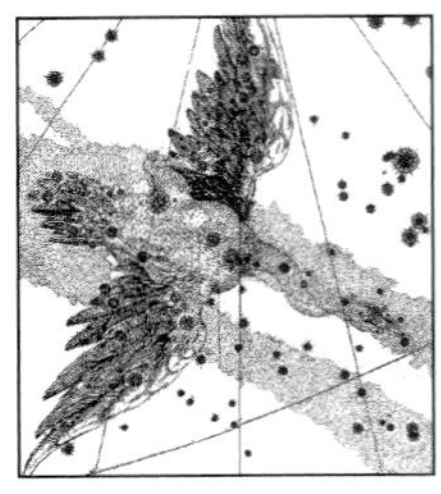

This 1822 star map of Cygnus shows the form of the swan that gives the constellation its name.

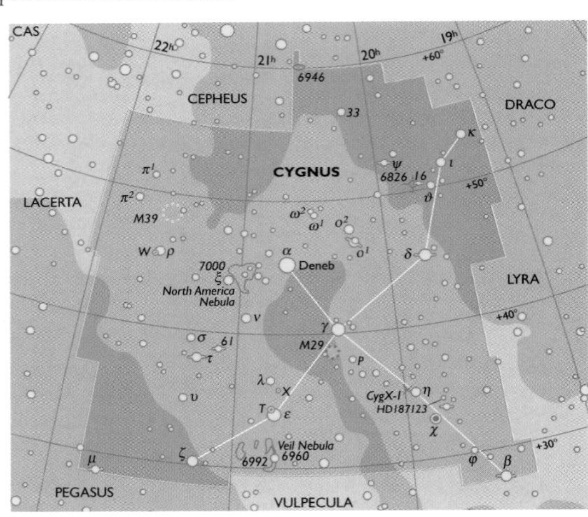

FEATURES OF INTEREST

ALPHA (α) CYGNI (DENEB): This is one of the most luminous stars in our galaxy, emitting more than sixty thousand times the light of our sun. Magnitude: +1.25; distance: 1,467 light-years.

BETA (β) CYGNI (ALBIREO): The most colorful double star in the heavens, a beautiful sight with any telescope. The primary is deep yellow, magnitude +3.08; the secondary is deep blue with a magnitude of +5.11. The separation is 34.5 arc seconds. The primary was revealed to be itself a double in 1977, although it is not resolvable with most amateur telescopes.

CHI (χ) CYGNI: A long-period variable star like Mira (Omicron Ceti) in the constellation Cetu, Chi Cygni varies in magnitude from +3.6—easily visible to the naked eye—to +14.2 over a 407-day period.

NGC 6992

This colorful nebulosity forms part of the eastern part of the Veil Nebula, a cloud of gas created when a nearby star went supernova and exploded.

NGC 7000
Also known as the North American Nebula for its resemblance to the continent, NGC 7000 stretches four moon diameters in the sky.

OMICRON 1 (ο1) CYGNI: A quadruple star, visible as a triple with steadily held binoculars. Component A: magnitude +4; component B: magnitude +7, with a separation of 107 arc seconds. Component C (30 Cygni) is a fifth-magnitude star that is 336 arc seconds distant.

61 CYGNI: At 11.36 light-years away, this is the thirteenth closest star to the sun. Sixty-one Cygni is a widely separated double star, easily resolved with a small telescope. The two orange dwarfs orbit each other every 650 years. Sixty-one Cygni was the first star to have its distance from Earth measured. In 1838, F. W. Bessel calculated the figure to within 10 percent of the current value.

P CYGNI: A peculiar variable star, possibly an ejection-type similar to Eta Carinae. Usually about magnitude +5, it can brighten or fade at any time.

CYGNUS X-1: This is an intense X-ray source associated with the unseen companion of a ninth magnitude, blue supergiant star. The companion is suspected to be a black hole that orbits the star every 5.6 days.

MESSIER 29: A small, coarse cluster of about a dozen magnitude 8 to 9 stars.

MESSIER 39: A large, scattered cluster of about twenty-five magnitude +7 and fainter stars.

NGC 6826: A remarkable planetary nebula with a fluorescent bluish tint. It is known as the Blinking Planetary for a peculiar visual feature. If you look directly at the nebula, you can see the magnitude +10.4 central star while the surrounding cloud seems to disappear. But if you look slightly away, the cloud overwhelms the star and it winks out. Do this rapidly and the nebula appears to blink on and off.

NGC 6960/6992: This is the Veil Nebula, beautiful in photographs and very well seen with moderate-aperture telescopes and an Oxygen III (OIII) filter. It is faintly visible in large, steadily held binoculars in very clear skies. The true nature of this object, and why it shines, is unknown. Most probably, the exploding power of a supernova tens of thousands of years ago swept up gas and dust in the interstellar medium, causing it to glow.

NGC 7000: The North American Nebula, so-named because its form resembles the terrestrial continent. The cloud is illuminated by Deneb (Alpha Cygni).

DELPHINUS (DEL-FIE-NUS) | *The Dolphin*

• **Abbreviation:** Del • **Genitive:** Delphini • **Area:** 189 square degrees • **Size Ranking:** 69th

• **Best Viewed:** August to September • **Latitude:** 90ºN–69ºS • **Width:** • **Height:**

Though fairly small and lacking in brilliant stars, this is an attractive naked-eye constellation. Both the Romans and Greeks saw the outline of a dolphin in Delphinus. In Greek mythology, Delphinus rescued the poet Arion from the hands of sailors who planned to kill him. The dolphin brought him ashore at Tarentum and Poseidon, in appreciation, set the dolphin among the stars. The constellation is also known as Job's Coffin, for a reason that escapes modern astronomers.

FEATURES OF INTEREST

GAMMA (γ) DELPHINI: A beautiful double star for the small telescope. The primary is magnitude +4.5; the secondary is fifth magnitude, separated by 10.1 arc seconds.

NGC 6891: A magnitude +10.5 planetary nebula with a strong bluish tint, appearing quite round in an eight-inch (20-cm) telescope.

NGC 6905: This magnitude +11.1 planetary nebula is about forty arc seconds in diameter.

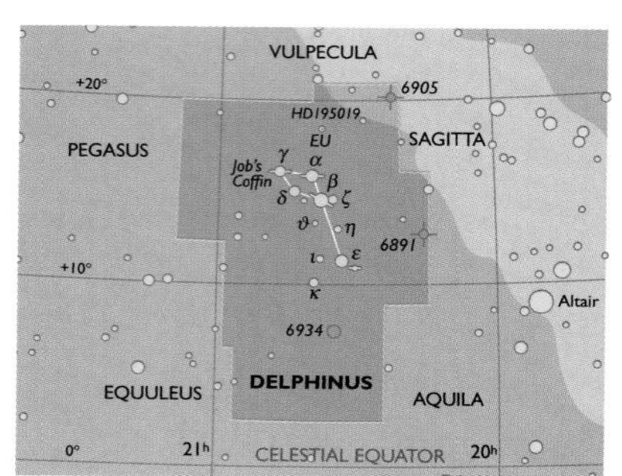

DORADO (DOH-RAH-DOH) | *The Swordfish*

• **Abbreviation:** Dor • **Genitive:** Doradus • **Area:** 179 square degrees • **Size Ranking:** 72nd

• **Best Viewed:** December to January • **Latitude:** 20ºN–90ºS • **Width:** • **Height:**

This small southern constellation, introduced by Bayer in 1604, represents a fish. Located well outside the boundaries of the Milky Way, a number of galaxies are visible here. The constellation's claim to fame is that much of the Large Magellanic Cloud, a satellite galaxy to our Milky Way, is located within its boundaries.

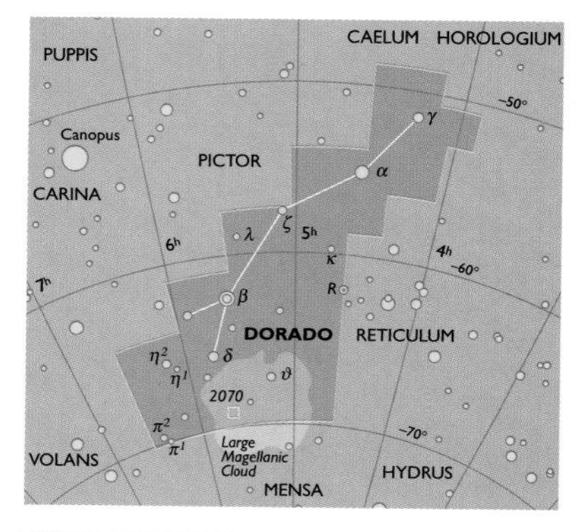

FEATURES OF INTEREST

THE LARGE MAGELLANIC CLOUD: The LMC was once classed as an irregular galaxy, but photographs often show evidence of a barred-spiral structure. As satellite galaxies go, it is huge: the central bar is at least twenty thousand light-years long. LMC is a very active galaxy, rich in star clusters and nebulae that can provide a season's worth of observing for an amateur with a large telescope. Within its confines are no less than 250 objects listed in the New General Catalog (NGC).

NGC 2070: A bright nebula and star cluster found within the Large Magellanic Cloud. This star-forming region is also known as the Tarantula Nebula.

NGC 2070

This enormous nebula is about eight hundred light-years in diameter. At its center lies a group of hot, young stars.

DRACO (DRAY-KO) | *The Dragon*

- **Abbreviation:** Dra • **Genitive:** Draconis • **Area:** 1,083 square degrees • **Size Ranking:** 8th
- **Best Viewed:** March to September • **Latitude:** 90ºN–4ºS • **Width:** • **Height:**

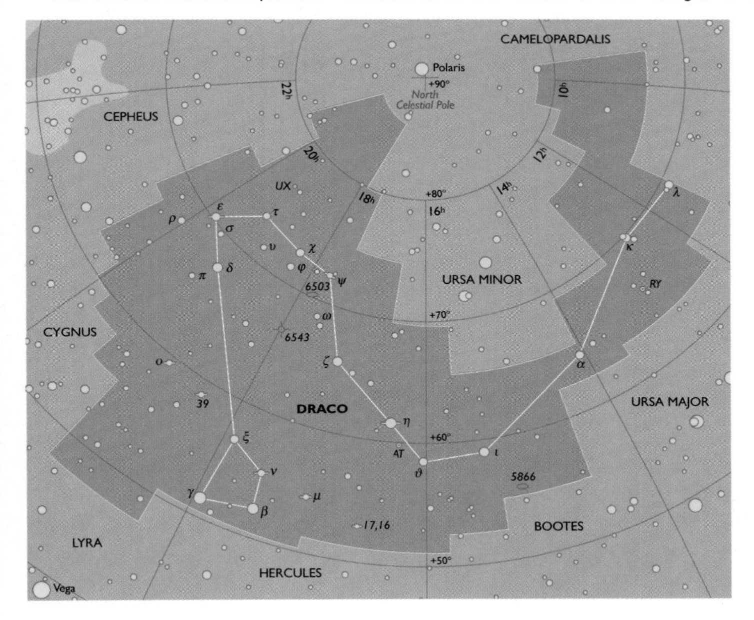

This extensive constellation snakes around the north polar region of the sky, with its head near the constellation Lyra and the end stars of the tail near the bowl of the Big Dipper. Many cultures identified this constellation as a dragon or a serpent. In Greek mythology, this is the dragon that lurked in the Garden of the Hesperides, ultimately slain by Hercules. For the patient explorer, it is an interesting region of the sky, featuring a bright planetary nebula and many galaxies.

FEATURES OF INTEREST

ALPHA (α) DRACONIS (THUBAN): Five thousand years ago, this was the Pole Star, coming closer to marking true north—within ten arc minutes—than Polaris does in the present age.

16-17 DRACONIS: A double star visible in binoculars (magnitudes +5 and +5.5), separated by ninety arc seconds.

NU (ν) DRACONIS: A double star visible in binoculars. Magnitudes: +4.95 and +4.98, separated by sixty-two arc seconds.

39 DRACONIS: A triple star: Component A: magnitude +5; component B: magnitude +8, separated by 3.7 arc seconds. Component C is magnitude +7.5, separated by eighty-nine arc seconds. Components A and C are visible in binoculars.

NGC 5866: Some obsolete references call this galaxy Messier 102, though it is now known that M101 and M102 were observations of the same galaxy. NGC 5866 is well worth seeking out. It is a bright, lens-shaped galaxy similar in form to NGC 3115 in Sextans.

NGC 6503: A spiral galaxy, NGC 6503 is easily visible with four-inch (10-cm) and larger apertures as an elongated nebula with a slightly brighter core.

NGC 6543: Also known as the Cat's Eye Nebula, this planetary nebula appears roundish and small in a small telescope. Though the central star is quite bright, it is difficult to detect owing to the brightness of the blue-green gaseous shell that surrounds it.

NGC 6543
Eighth-magnitude planetary nebula NGC 6543 is a gaseous cloud that formed a thousand years ago from a central star.

EQUULEUS (EH-KWOO-LEE-US) | *The Little Horse*

• **Abbreviation:** Equ • **Genitive:** Equulei • **Area:** 72 square degrees • **Size Ranking:** 87th

• **Best Viewed:** September • **Latitude:** 90ºN–4ºS • **Width:** ✋ • **Height:** ✋

This small, inconspicuous constellation is difficult to identify without a star atlas handy. To the ancient Greeks, the Little Horse represented Cyllarus, the horse given to Pollux by Hera. A few faint galaxies are located here. For the small-telescope owner, there are a couple of multiple stars of interest.

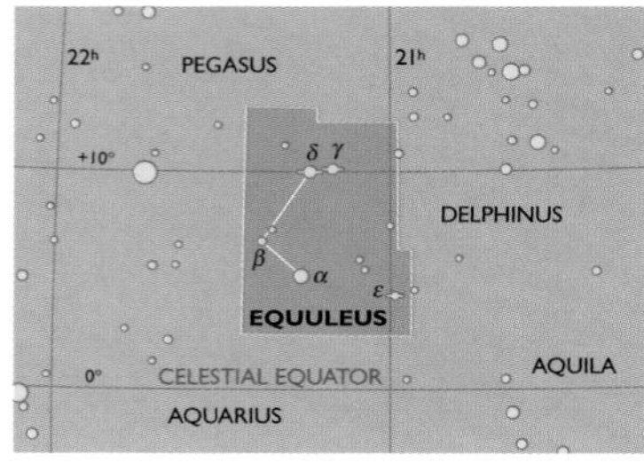

In Greek mythology, Equuleus was the brother of the winged horse Pegasus. The constellation is the second smallest one in the night sky.

FEATURES OF INTEREST

GAMMA (γ) EQUULEI: Along with Epsilon (ε) Equulei, Gamma (γ) Equulei forms a wide binocular pair. Gamma itself is a triple, with a magnitude +11 companion 1.9 arc seconds distant and another magnitude +12 star about 48 arc seconds away.

EPSILON (ε) EQUULEI: A triple-star system. The A and B components are separated by less than 0.8 arc seconds and require a large amateur telescope to split. A third component, magnitude +7, is located 10.6 arc seconds away.

ERIDANUS (EH-RID-AN-US) | *The River*

• **Abbreviation:** Eri • **Genitive:** Eridani • **Area:** 1,138 square degrees • **Size Ranking:** 6th

• **Best Viewed:** November to January • **Latitude:** 32ºN–89ºS • **Width:** ✋✋ • **Height:** ✋✋✋

Eridanus is a long, meandering constellation that all ancient cultures identified with a river—usually the one that was central to their lives. The constellation extends more than fifty-seven degrees from north to south and is the sixth largest by area. It contains one first-magnitude star as well as many double stars and is a rich hunting ground for galaxies, although some of them require a twelve-inch (30-cm) telescope to see.

FEATURES OF INTEREST

ALPHA (α) ERIDANI (ARCHERNAR): This blue-white star, magnitude 0.5, owes its name to the Arabic word meaning "river's end." Archenar marks the southern extremity of the constellation.

EPSILON (ε) ERIDANI: At a distance of 10.5 light-years, this star ranks as the thirteenth closest to the sun. Magnitude: +3.72.

THETA (ϑ) ERIDANI: An excellent double star for the small telescope, this white pair, magnitudes +3.5 and +4.5, is separated by 8.2 arc seconds.

OMICRON 2(o2) ERIDANI: This is a remarkable triple-star system, a little too faint for binoculars. Component A is magnitude +4.43. Components B and C are 82.8 arc seconds distant, a binary pair that orbit each other every 248 years. Component B—the easiest to see with a small telescope—is a magnitude 9.7 white dwarf while component C is a red dwarf, a very rare combination. Magnitude 10.8.

32 ERIDANI: A double star of different-colored stars. Component A: magnitude +5; component B: magnitude +6. The two are separated by 6.9 arc seconds.

NGC 1084: A small spiral galaxy that is easily seen in six-inch (15-cm) and larger telescopes as an oval glow oriented northeast/southwest. Magnitude: +11.2.

NGC 1232: Photographs show this as a fine multiple-arm spiral galaxy. It appears oval with a brighter core, slightly elongated east/west. Magnitude: +10.5.

NGC 1300: A classic barred-spiral galaxy, but visually disappointing in a small telescope since the spiral arms are too faint to see. The core is prominent, and patience and superb skies will show faint extensions east and west. Magnitude +11.1.

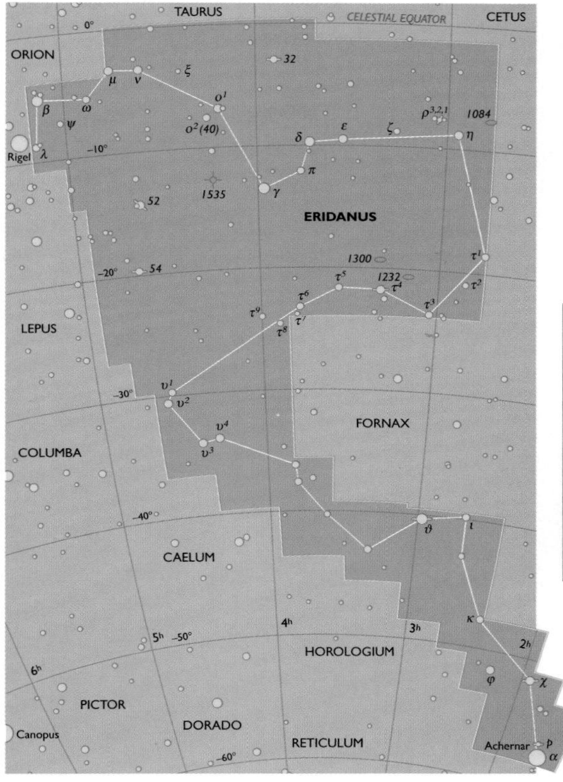

NGC 1535:
A fine planetary nebula that is about thirty arc seconds in diameter with a magnitude of 9.6. As with many of the brighter examples of planetary nebula, there is a strong bluish tint and a grainy fluorescence to the cloud of dust and gas. The central star that illuminates NGC 1535 is visible.

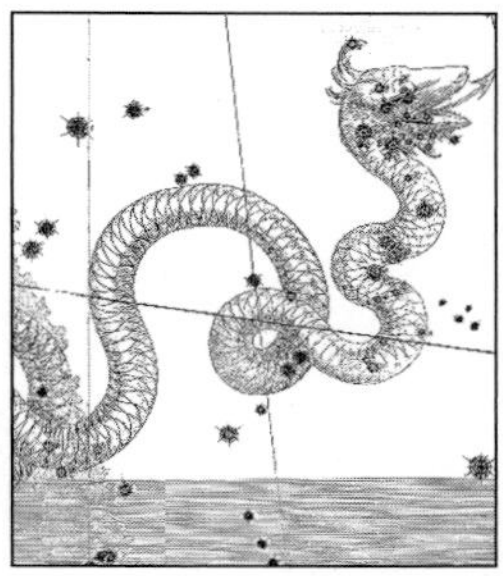

In Johann Bayer's Uranometria *star atlas of 1603, Eridanus the River is shown as a serpent with the magnitude 0.5 star Achernar at its mouth.*

FORNAX (FOR-NAX) | *The Furnace*

- **Abbreviation:** For • **Genitive:** Fornacis • **Area:** 398 square degrees • **Size Ranking:** 41st
- **Best Viewed:** November to December • **Latitude:** 50ºN–90ºS • **Width:** • **Height:**

Although not far south, Fornax was not created until 1752, when French astronomer Lacaille delineated it. No star is brighter than the fourth magnitude, and at first glance there is little to distinguish it. But it is home to a rich galaxy group, suitable for eight-inch (20-cm) and larger telescopes.

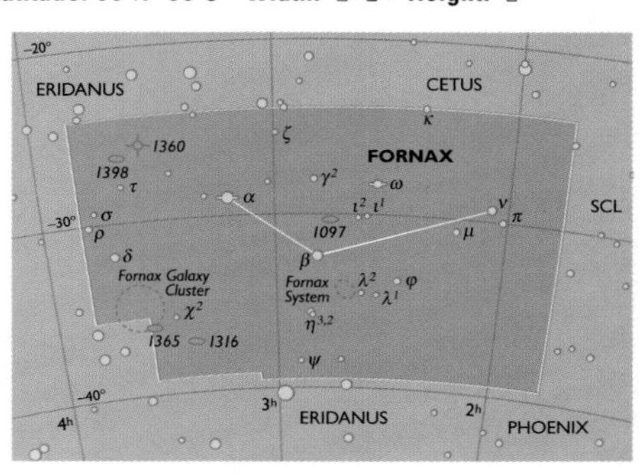

FEATURES OF INTEREST

FORNAX GALAXY CLUSTER: At least eighteen galaxies should be detectable in this cluster. While the brightest is NGC 1316 at magnitude +10, the most visually impressive is the classic barred-spiral NGC 1365.

NGC 1097: A typical barred-spiral galaxy with extremely faint outer spiral arms. Magnitude: +10.2.

GEMINI (JEM-EH-NYE) | *The Twins*

• **Abbreviation:** Gem • **Genitive:** Geminorum • **Area:** 514 square degrees • **Size Ranking:** 30th

• **Best Viewed:** January to February • **Latitude:** 90ºN–55ºS • **Width:** • **Height:**

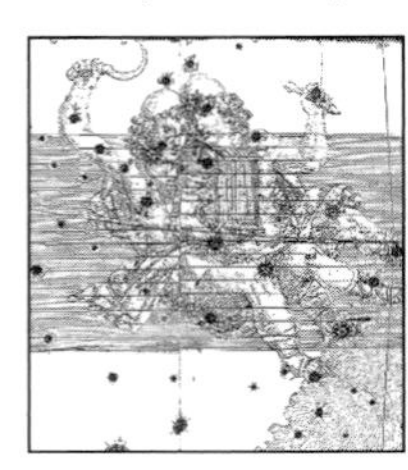

An 1822 star atlas shows the mythological twins Castor and Pollux.

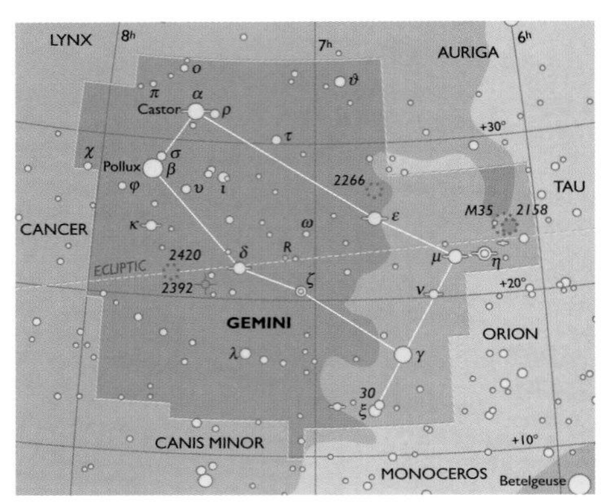

This constellation is a large, fairly bright assemblage of stars featuring the twin first-magnitude luminaries Castor and Pollux. All ancient cultures associated this constellation with twins, usually men but sometimes plants or animals. In classical mythology, Castor and Pollux were the illegitimate children of Leda, queen of Sparta, who was pursued and eventually seduced by Zeus. The twins were among the crew of Jason's Argonauts and protected his ship and its crew from the dangers of the sea. In Roman mythology, Castor was mortal while Pollux was immortal. They represented true brotherly love, and Pollux would sometimes cede his place in heaven to his mortal brother. With the feet of the twins immersed in the Milky Way, this is an interesting region of the sky to explore with binoculars or a small telescope. One Messier object is located here, the beautiful open cluster M35.

FEATURES OF INTEREST

ALPHA (α) GEMINORUM (CASTOR): This is a remarkable six-star family, though only three of the stars are visible in telescopes. Each one of the visible stars is, in fact, a binary star with components too close together to be detected visually. Castor A is magnitude +1.94; both it and its companion are stars of similar mass and brightness. Castor B is located 3.9 arc seconds distant and again, both components are almost identical, with a magnitude of +2.92. Castor C, located 72.5 arc seconds distant, is a red dwarf pair, magnitude +9. The orbital period of A and B is about 470 years, while the period of component C is unknown.

BETA(β) GEMINORUM (POLLUX): At magnitude +1.14, this orange giant is slightly brighter than Castor. It is located thirty-four light-years away—seventeen light-years closer than Castor.

ZETA (ζ) GEMINORUM: A Cepheid variable—a pulsating star with a brightness that changes over time. In the case of Zeta Geminorum, the period is 10.15 days, during which the star varies between magnitudes +3.6 and +4.2.

The constellation is dominated by Castor and Pollux. In this photograph, Castor appears in the upper center left, while Pollux lies below it to the left.

ETA (η) GEMINORUM: A long-period variable, with a semi-regular period of 233 days. Its brightness ranges from +3.1 to +3.9. It is also a double star, but very difficult to split for the small telescope since the companion is magnitude +8.8 and separated by only 1.6 arc seconds. The planet Uranus was discovered near this star by William Herschel in 1781.

MESSIER 35: An excellent star cluster for the small telescope, M35 can be detected with the naked eye under pristine dark-sky conditions and is an easy object for binoculars. Owing to its large size (about thirty arc minutes in diameter), it is best seen at low magnification: a very rich cluster featuring prominent chains of stars. It lies about twenty-three hundred light-years distant. Four-inch (10-cm) and larger telescopes should have no problem locating a misty patch near the cluster's southwestern extremity. This is NGC 2158, one of the richest open clusters in our galaxy, but extremely distant (about 16,800 light-years).

NGC 2266: This is a rich cluster, although difficult to resolve with smaller apertures. With an eight-inch (20-cm) telescope, it appears arrow-shaped with a magnitude +9 star at the tip.

NGC 2392: This planetary nebula is also known as the Clown Face and Eskimo Nebula, although the features that gave it these nicknames are visible only at high magnification with twelve-inch (30-cm) and larger telescopes. NGC 2392 appears as a blue-green disk, with a bright central star. Magnitude: +8.3; diameter: about thirty arc seconds.

NGC 2420: An open cluster located in a rich star field: a hazy V-shaped nebulous haze with several stars that can be seen in an eight-inch (20-cm) telescope.

GEMINIDS: One of the finest, most dependable meteor showers. Its maximum display is on December 13th or 14th. Expect to see up sixty meteors per hour in a moonless sky.

GRUS (GROOS) | *The Crane*

• **Abbreviation:** Gru • **Genitive:** Gruis • **Area:** 366 square degrees • **Size Ranking:** 45th

• **Best Viewed:** September to October • **Latitude:** 33ºN–90ºS • **Width:** • **Height:**

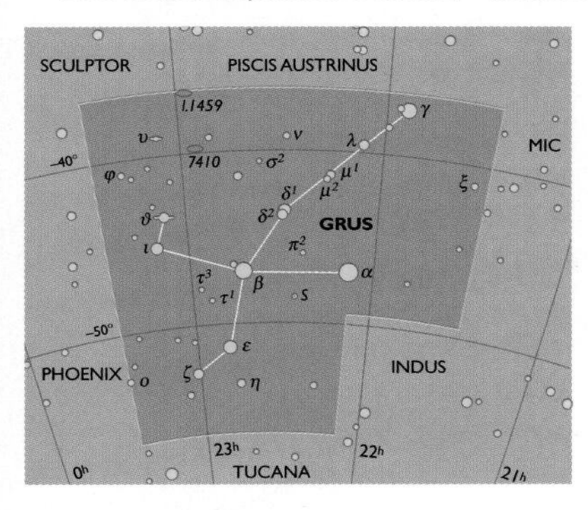

Located far from the plane of the Milky Way, Grus is an excellent hunting ground for other galaxies, although the majority are quite faint. The constellation is moderately bright and easy to identify, but a little too far south to be well seen by mid-northern observers. It was first delineated by Johann Bayer in his star atlas *Uranometria* in 1603.

FEATURES OF INTEREST

ALPHA (α) GRUIS (ALNAIR): A large blue star, magnitude 1.7, with a faint companion 28.3 arc seconds distant. The moderately bright galaxy NGC 7213 is located only sixteen arc minutes to the southeast.

DELTA 1 (δ1) AND DELTA 2 (δ2) GRUIS: A double star three hundred light-years distant that can easily be separated with binoculars.

MU 1 (μ1) AND MU 2 (μ2) GRUIS: A widely spaced double star, magnitudes: 4.8 and 5.1.

NGC 7410: This moderately bright (magnitude +11.3) spiral galaxy features a bright core and tapered ends that are detectable with a small telescope.

IC 1459: A bright elliptical galaxy, slightly elongated northeast/southwest.

HERCULES (HER-KYU-LEEZ) | *The Strongman*

• **Abbreviation:** Her • **Genitive:** Herculis • **Area:** 1,225 square degrees • **Size Ranking:** 5th

• **Best Viewed:** June to August • **Latitude:** 90ºN–38ºS • **Width:** • **Height:**

Hercules is one of the most famous of mythological heroes, playing a prominent role in many of the stories of Greek mythology. He sailed with Jason and the Argonauts during their quest for the Golden Fleece and his legendary strength was immortalized in the "Twelve Labors of Hercules," including battles with the Nemaen Lion, the Stymphalian Birds, and the Cretan Bull. During his Tenth Labor, he created the Pillars of Hercules at the Straits of Gibraltar, which in ancient times marked the edge of the known world. The great-grandson of Perseus, Hercules was born of the union of Zeus and the mortal woman Alceme. The constellation named in his honor is made up primarily of third-magnitude stars. It is a distinctive pattern in the nighttime sky, well-known among amateur astronomers using large telescopes to search for distant galaxies and rich galaxy clusters. It is also an interesting area of the sky for the beginning observer with more modest equipment. Two Messier objects can be found here, the spectacular globular clusters M13 and M92.

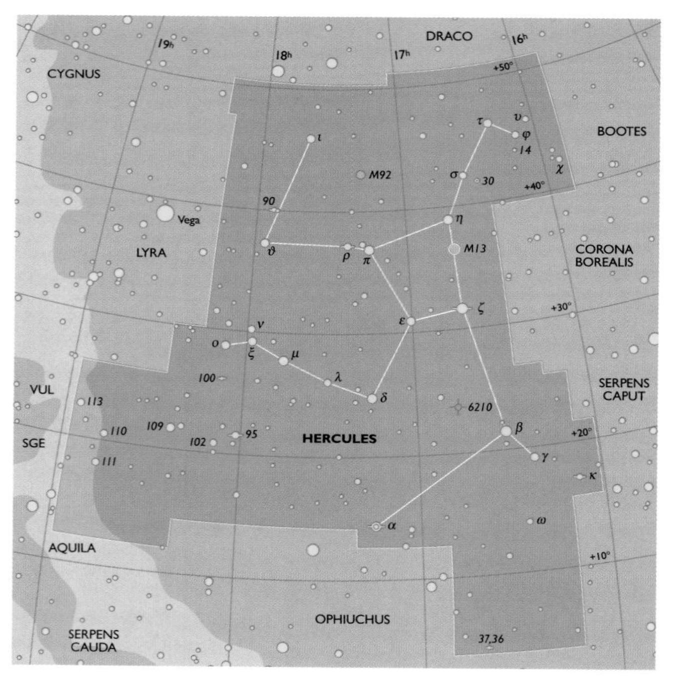

FEATURES OF INTEREST

ALPHA (α) HERCULIS (RAS ALGETHI): A bright, irregular variable star, reddish or orange in color, with a magnitude varying from +3 to +4 in a period ranging from 50 to 130 days. One of the largest known stars, this red supergiant is also a beautiful double. The magnitude +5.4 secondary is located 4.7 arc seconds from the primary.

ZETA (ζ) HERCULIS: A well-known double star, difficult to separate with a small telescope, but notable because of its relatively quick rotation period of thirty-four years. The current separation between the stars is 0.8 arc seconds.

95 HERCULIS: A pair of stars, silver and gold in color, with magnitudes 5.0 and 5.2.

MESSIER 13: The largest, brightest, and most spectacular globular cluster north of the celestial equator, outclassed only by Omega Centauri, 47 Tucanae, and Messier 22 in the southern sky. This is an extraordinary object, easily visible with binoculars, and a fine object for the small telescope.

Partial resolution of the extremities is possible with a three-inch (8-cm) aperture in dark skies; with eight-inch (20-cm) and larger telescopes, it is spectacular, resolving into thousands of stars. M13 is notable for a curious crab-like outline defined by the outlying stars. The diameter of the cluster is about twenty arc minutes, with a magnitude of +5.9. A galaxy, NGC 6207, is located northeast.

MESSIER 92: An interesting rival to M13, M92 is smaller and a little fainter, but beautiful nonetheless. It features a blazing core, much brighter than that of M13. The resolution of outlying stars can be made with a four-inch (10-cm) telescope because there are many bright stars at the extremities. Diameter: about fifteen arc minutes; magnitude: +6.4; distance: twenty-six thousand light-years.

NGC 6210: A very bright planetary nebula, with a strong bluish color. Magnitude: +9.3; diameter: 14 arc seconds.

M13
To the naked eye, M13 looks like a hazy star. But with a telescope, its thousands of stellar members become visible.

HOROLOGIUM (HOR-OH-LOH-JEE-UM) | *The Clock*

• **Abbreviation:** Hor • **Genitive:** Horologii • **Area:** 249 square degrees • **Size Ranking:** 58th

• **Best Viewed:** November to December • **Latitude:** 23ºN–90ºS • **Width:** • **Height:**

This rather dim southern constellation, representing a clock, was created by the French astronomer Nicolas de Lacaille in 1752. It is far from the plane of the Milky Way, and there are only a few deep-sky objects here of interest to the binocular and small-telescope user.

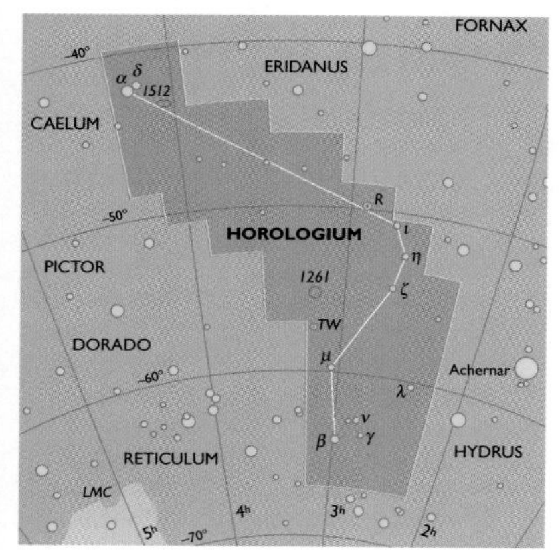

FEATURES OF INTEREST

R HOROLOGII: This long-period variable star brightens and fades from the fifth to the fourteenth magnitudes every thirteen months.

NGC 1261: This is a moderately bright globular star cluster, about 2.5 arc minutes across, with a very compressed structure. The component stars are very faint and require at least a twelve-inch (30-cm) telescope to resolve across the face of the cluster. Magnitude: +8; distance: about fifty-two thousand light-years.

NGC 1512: A barred-spiral galaxy that is quite faint with apertures less than six inches (15 cm).

HYDRA (HY-DRA) | *The Sea Serpent*

• **Abbreviation:** Hya • **Genitive:** Hydrae • **Area:** 1,303 square degrees • **Size Ranking:** 1st

• **Best Viewed:** January to February • **Latitude:** 54ºN–83ºS • **Width:** • **Height:**

This long and meandering constellation has the distinction of being the largest in the sky, stretching across a quarter of the sky. In Greek mythology, it was a nine-headed water snake. If one of its heads was cut off, it immediately grew back. Eventually, Hercules slew the monster with the aid of Iolas, who cauterized each severed neck with a branding iron. The head of this constellation is marked by a prominent loop of six stars almost due west of the first-magnitude star Regulus. There are many interesting objects here, including three Messier objects: M48, M68, and M83.

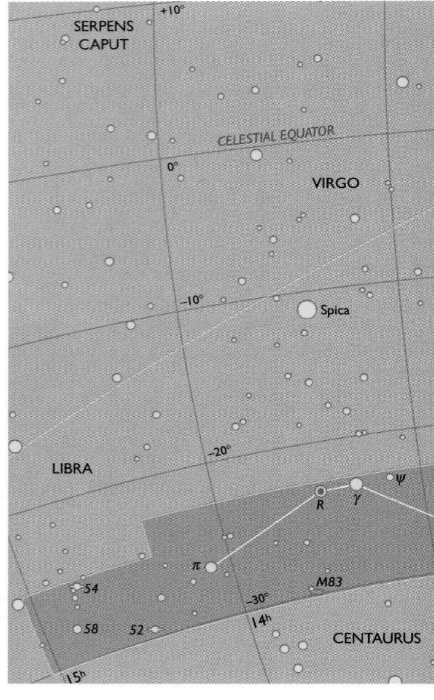

M83
One of the most prominent of the barred-spiral galaxies, M83 has been the birthplace of four supernovas since the 1930s.

FEATURES OF INTEREST

ALPHA (α) HYDRAE (ALPHARD): An orange-colored giant, magnitude 2, known as the Solitary One, owing to its isolated position in the sky.

EPSILON (ε) HYDRAE: A four-star system, magnitude +3.36, with three stars visible with a small telescope. The primary is a very tight double with an orbital period of fifteen years, resolvable only with very large telescopes. The third component is magnitude +8 and located 3.1 arc seconds away. The fourth component is 19.3 arc seconds distant and magnitude +12.

R HYDRAE: A variable, Mira-type star, which ranges from +4 to +10 magnitude every 386 days.

U HYDRAE: A variable star, reddish in color, that fluctuates in magnitude from +4.7 to +6.2.

MESSIER 48: One of the "missing" Messier objects, owing to an error in position made by Charles Messier when he compiled his famous catalog. In fact, it is the same object as NGC 2548. This is a beautiful cluster for binoculars, which can resolve individual stars. Magnitude: +5.8.

MESSIER 83: Situated twenty-two million light-years away, M83 is an impressive spiral galaxy, with thick, bright spiral arms. From mid-northern latitudes, it usually appears as a pale, roundish object with a brighter core, but under exceptional conditions the spiral pattern can be seen well with an eight-inch (20-cm) telescope. Magnitude: +8.51.

NGC 3242: This planetary nebula is also called the Ghost of Jupiter for its pale resemblance to the king of planets. Although the nebula can be detected with even the smallest telescope, its bright elliptical inner region and pale surrounding outer shell require at least a twelve-inch (30-cm) telescope to be seen properly. Magnitude: +8.6.

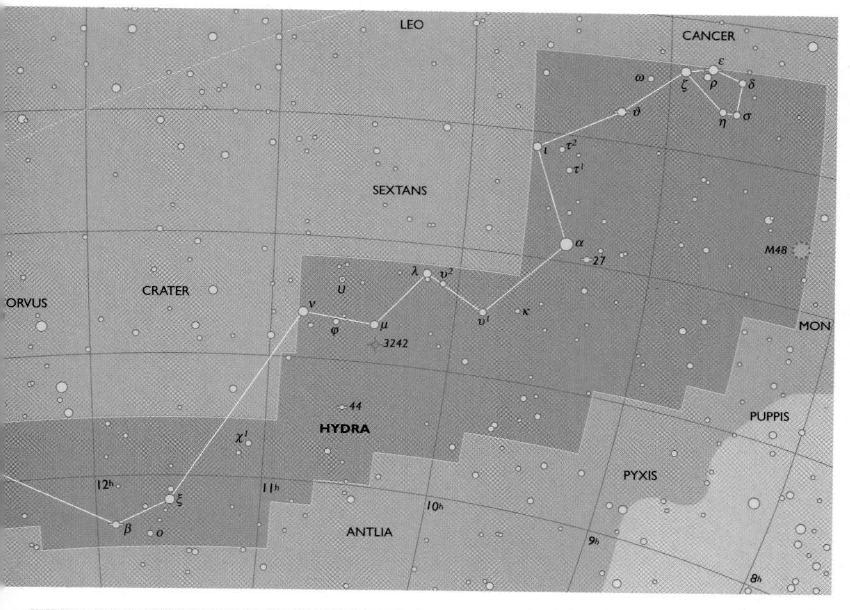

HYDRUS (HIGH-DRUS) | *The Water Snake*

• **Abbreviation:** Hyi • **Genitive:** Hydri • **Area:** 243 square degrees • **Size Ranking:** 61st

• **Best Viewed:** October to December • **Latitude:** 8ºN–90ºS • **Width:** • **Height:**

The far-southern constellation Hydrus is located between the Large and Small Magellanic Clouds, irregular galaxies that are satellites of the Milky Way. The constellation was defined by two Dutch navigators in the sixteenth century, then included in Johann Bayer's star atlas of 1603. The three brightest stars form a prominent triangular outline pointing almost directly opposite the south celestial pole.

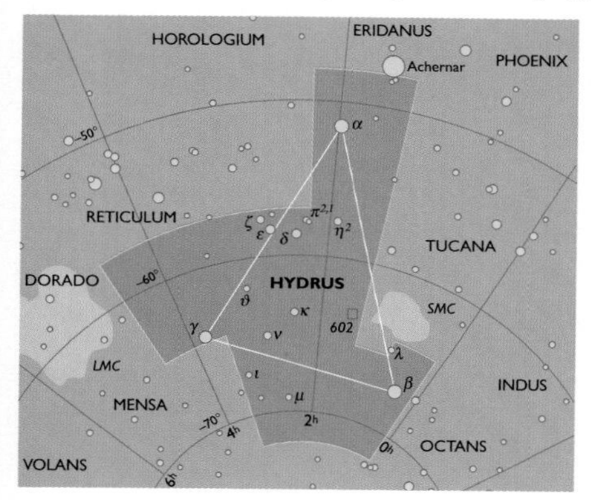

FEATURES OF INTEREST

PI (π) HYDRI: Fifth-magnitude double stars related only by their proximity in the sky. Pi-1 lies about 740 light-years away; Pi-2 is 470 light-years distant.

NGC 602: This gaseous nebula is associated with the Small Magellanic Cloud. Easily seen with a six-inch (15-cm) telescope, NGC 602 measures about 0.7 by 1.5 arc minutes with a dark lane bisecting the bright nebula.

INDUS (IN-DUS) | *The Indian*

• **Abbreviation:** Ind • **Genitive:** Indi • **Area:** 294 square degrees • **Size Ranking:** 49th

• **Best Viewed:** August to October • **Latitude:** 15ºN–90ºS • **Width:** • **Height:**

Another deep-southern constellation included in Johannes Bayer's 1603 star atlas, Indus represents a North American Indian. Except for Alpha (α) Indi, the stars are quite faint. There is a good selection of galaxies here for telescopes eight inches (20 cm) and larger; these objects are all located in the northern reaches of the constellation.

FEATURES OF INTEREST

EPSILON (ε) INDI: Similar in size to our sun, Epsilon Indi has been investigated by radio astronomers searching for signs of planets and extraterrestial life. This magnitude +4.68 star is located 11.83 light-years from Earth.

THETA (ϑ) INDI: A double star with 4.5- and 6.9-magnitude components, separated by seven arc seconds.

IC 5152: An irregular, 11.6-magnitude galaxy, with an eighth-magnitude star in its field that outshines the galaxy, making IC 5152 difficult to view.

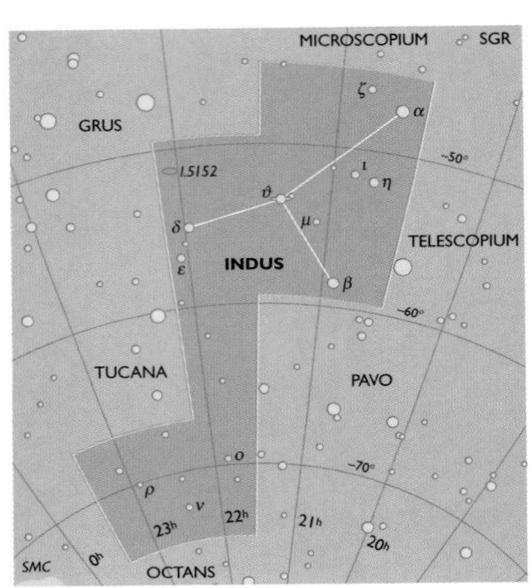

LACERTA (LAH-SIR-TAH) | *The Lizard*

• **Abbreviation:** Lac • **Genitive:** Lacertae • **Area:** 201 square degrees • **Size Ranking:** 68th

• **Best Viewed:** September to October • **Latitude:** 90ºN–33ºS • **Width:** • **Height:**

Lacerta the Lizard was created out of a collection of faint stars in the northern Milky Way by Johannes Hevelius in 1687. There are a handful of fairly bright, open star clusters in this region.

FEATURES OF INTEREST

BL LACERTAE: Originally classed as a variable star, this is, in fact, a very distant galaxy with a faint or invisible outer envelope. The core is usually visible as a star-like point of light, ranging between magnitude +14 and +17, making it an object of study for very large amateur telescopes. Scientists speculate that at the heart of BL Lacertae-type (BL Lac) objects lies a black hole.

NGC 7209: A fairly large, scattered cluster of magnitude +8 and fainter stars in a rich, Milky Way field. About forty to fifty stars are visible with an eight-inch (20-cm) telescope.

NGC 7243: A more widely scattered cluster than NGC 7209, both in terms of size and brightness of cluster members. A conspicuous double star, both members about magnitude +9, lies near the heart of the cluster. An eight-inch (20-cm) telescope will show about forty to fifty stars.

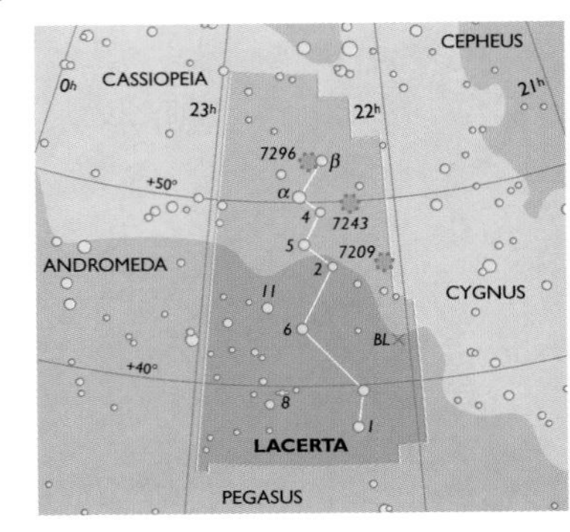

LEO (LEE-OH) | *The Lion*

• **Abbreviation:** Leo • **Genitive:** Leonis • **Area:** 947 square degrees • **Size Ranking:** 12th
• **Best Viewed:** March to April • **Latitude:** 82ºN–57ºS • **Width:** • **Height:**

This large, bright constellation is one of the few in the sky that looks something like the form it is supposed to represent: a crouching lion. Some historians have suggested that the form of the Egyptian sphinx was based on this heavenly counterpart. Located well away from the plane of the Milky Way, the constellation subsumes a very rich area of the sky for galaxies, with dozens of bright examples for the small-telescope user.

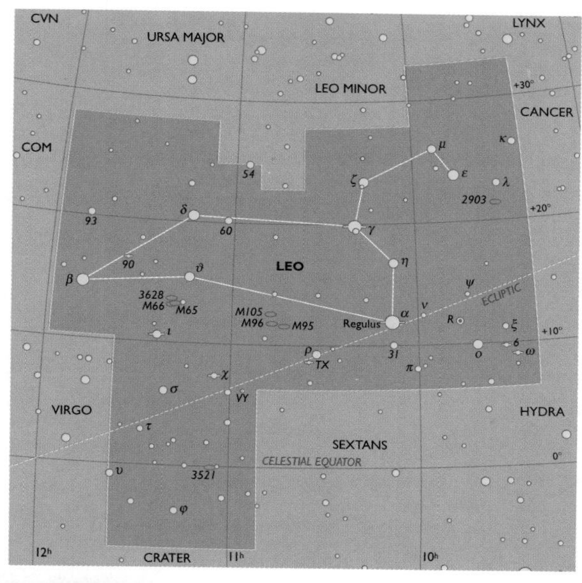

M66
Situated forty million light-years from Earth, M66 is a spiral galaxy that appears elliptical because of its tilt.

FEATURES OF INTEREST

LEONIDS: One of the most prominent of annual meteor showers, the Leonids last several days, reaching their peak every year on November 17th.

ALPHA (α) LEONIS (REGULUS): A blue-white star, magnitude +1.35. There are at least three faint companions visible near it, although they are unrelated to Regulus.

GAMMA (γ) LEONIS: A brilliant double star, primary magnitude +2.61, secondary magnitude +3.47. The two, which are yellow and orange, are separated by 4.4 arc seconds, a gap that is very slowly widening. Immediately east is the faint galaxy pair NGC 3226/3227.

R LEONIS: A variable red star of the Mira-class, with magnitudes ranging from +5.2 to +10.5 in a period of 313 days.

MESSIER 65, MESSIER 66, NGC 3628: A triple-galaxy system visible together in the low magnification field of a small telescope. M65 is a spiral galaxy, magnitude +9.3. M66, another spiral, is slightly brighter at magnitude +9. Immediately north of these two lies the slightly fainter NGC 3628, a large galaxy that is seen edge on with a dust lane bisecting it, which can be faintly detected with fourteen-inch (36-cm) and larger telescopes.

MESSIER 95, MESSIER 96: M95 is a dusty barred-spiral galaxy with a faint encircling spiral ring, unfortunately not visible with a small telescope. M96 is a spiral galaxy with a very large, bright core.

MESSIER 105: Bright (magnitude +9.3) elliptical galaxy, forming a triple system in a low-power eyepiece with NGC 3384 and NGC 3389. All three galaxies are visible with a six-inch (15-cm) telescope under good conditions.

NGC 2903: Very large and bright spiral galaxy, a good example of Hubble's Sb-class galaxy, a type with dusty, thick spiral arms. Magnitude: +8.9.

LEO MINOR (LEE-OH MY-NOR) | *The Little Lion*

• **Abbreviation:** LMi • **Genitive:** Leonis Minoris • **Area:** 232 square degrees • **Size Ranking:** 64th

• **Best Viewed:** March to April • **Latitude:** 90ºN–48ºS • **Width:** • **Height:**

This small, faint constellation, added by Hevelius in the year 1660, contains no star brighter than magnitude +4. Located immediately north of its larger, brighter namesake, it represents a lion cub.

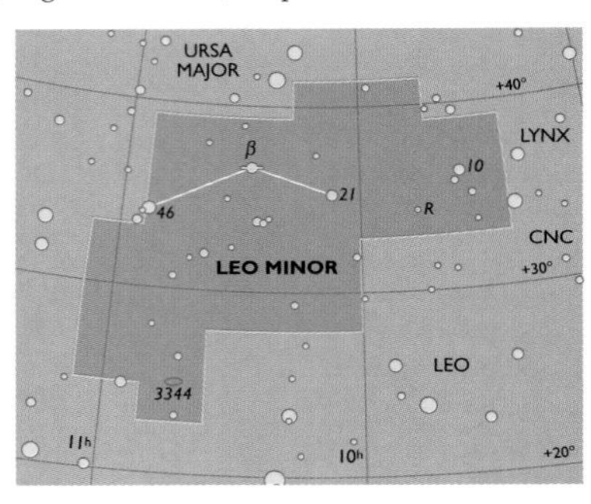

FEATURES OF INTEREST

BETA (β) LEONIS: A very close double star, with components of magnitude +4.4 and +6.1. The two orbit around each other every thirty-seven years. With anything less than a fourteen-inch (36-cm) telescope, this will appear as a single star. Beta Leonis is the only star in the constellation with a Greek letter. All the others have numbers assigned by the English astronomer John Flamsteed in the eighteenth century.

NGC 3344: A diffuse, face-on spiral galaxy, magnitude +9.3. This is a fairly bright specimen. The outer spiral arms are visible as a faint glow surrounding a bright core.

LEPUS (LEE-PUS) | *The Hare*

• **Abbreviation:** Lep • **Genitive:** Leporis • **Area:** 290 square degrees • **Size Ranking:** 51st

• **Best Viewed:** January • **Latitude:** 62ºN–90ºS • **Width:** • **Height:**

Lepus the Hare, one of Ptolemy's original forty-eight constellations, is identified with a rabbit in Roman mythology and was placed in the sky because Orion liked hunting them. This is a conspicuous grouping located immediately south of Orion. It contains a number of interesting deep-sky objects, including the globular cluster M79.

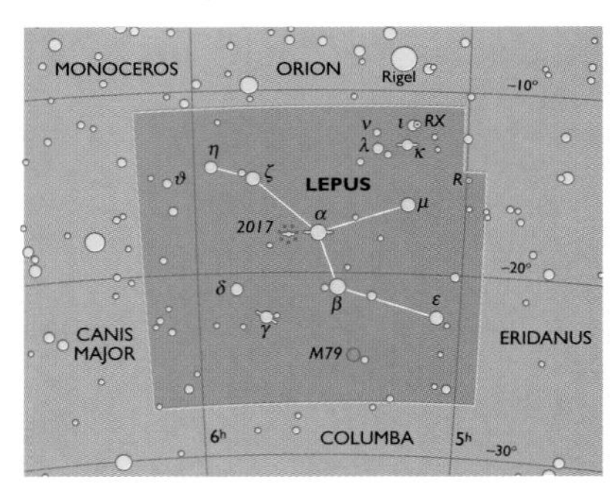

FEATURES OF INTEREST

ALPHA (α) LEPORIS (ARNEB): This star has two faint unrelated companions: one magnitude +11, separated by 35.3 arc seconds; the other magnitude +12, separated by 91.4 arc seconds.

GAMMA (γ) LEPORIS: A double star suitable for steadily held binoculars. Component A is magnitude +3.6; the secondary is magnitude +6.18, separated by 1.5 arc minutes. The stars are contrasting colors of orange and yellow. There is a third, unrelated star, magnitude +11, forty-five arc seconds distant from component B.

R LEPORIS: A Mira-type variable star that ranges from +5.9 to fainter than +11 magnitude over a period of 432 days. The nineteenth-century British astronomer Russel Hind dubbed it the Crimson Star. And for good reason: Its color has been likened to an illuminated drop of blood.

MESSIER 79: A fairly typical globular cluster in terms of size and brightness, this is an easily detected object in a small telescope with its bright and compressed core region. With eight-inch (20-cm) and larger telescopes, you should be able to resolve magnitude +14 and fainter outlying stars.

NGC 2017: This multiple star system is also cataloged as h3780. Small telescopes will reveal six stars here, with a wide brightness range, well isolated from the sky background.

LIBRA (LEE-BRA) | *The Scales*

• **Abbreviation:** Lib • **Genitive:**Librae • **Area:** 538 square degrees • **Size Ranking:** 29th
• **Best Viewed:** May to June • **Latitude:** 60ºN–90ºS • **Width:** • **Height:**

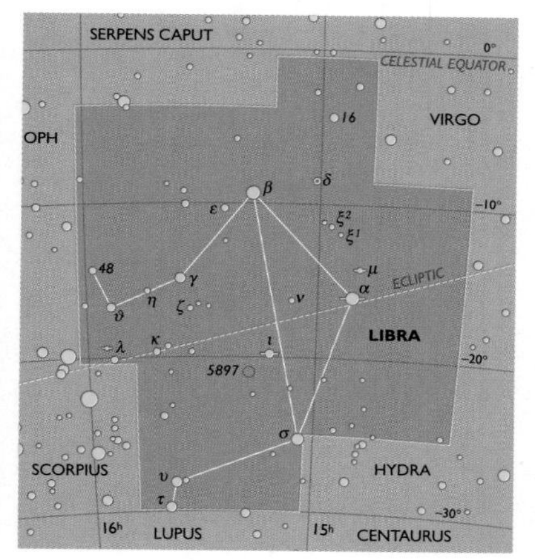

Though this star grouping has been known since ancient times, Ptolemy did not represent it as a separate constellation. For him, it formed an extension of Scorpius, delineating the claws of the scorpion: Alpha (α) Librae (Zubenelgenubi) being the northern claw, and Beta (β) Librae (Zubeneschamali) the southern claw. Long known as the Scales, the symbol of justice, Libra features more than a dozen galaxies suitable for eight-inch (20-cm) and larger instruments.

FEATURES OF INTEREST

DELTA (δ) LIBRAE: Like Algol in the constellation Perseus, this is an eclipsing variable star, in which the two stars regularly pass in front of each other, causing the magnitude to vary—in this case, between between +4.8 and +5.9 every 2.3 days.

NGC 5897: A loose globular cluster; eight-inch (20-cm) telescope needed to see it faintly.

LUPUS (LOO-PUS) | *The Wolf*

• **Abbreviation:** Lup • **Genitive:** Lupus • **Area:** 334 square degrees • **Size Ranking:** 46th
• **Best Viewed:** May to June • **Latitude:** 60ºN–90ºS • **Width:** • **Height:**

Lupus the Wolf is one of Ptolemy's forty-eight constellations and is located in a rich region of the Milky Way, between Scorpius and Centaurus. There is a wide variety of star clusters and nebulae here for the small telescope.

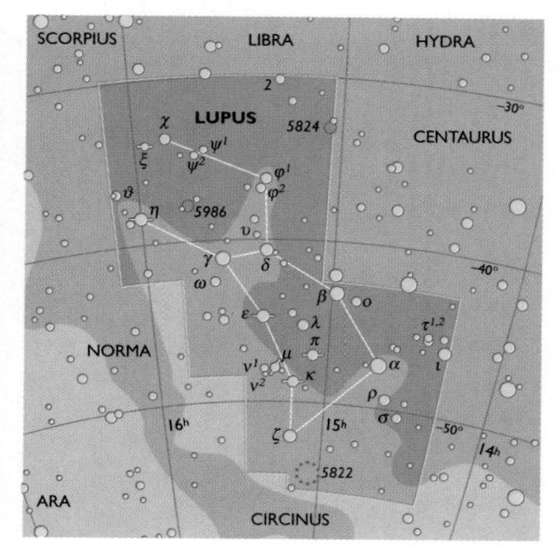

FEATURES OF INTEREST

ALPHA (α) LUPI: A variable star that changes very slightly in brightness every 0.26 days.

KAPPA (κ) LUPI: A double star, magnitudes +4 and +6, separated by twenty-seven arc seconds.

PI (π) LUPI: Visible with a three-inch (8-cm) telescope, this double-star system consists of two blue-white stars of the same magnitude (+5) separated by 1.4 arc seconds.

XI (ξ) lupi: A double star suitable for a small telescope user. Both components are magnitude +5.5, separated by 10.6 arc seconds.

NGC 5822: A very large, extensive cluster of stars. It is best seen with a low-power, wide-field eyepiece. About a hundred stars are visible, ranging from magnitudes +9 to +12.

NGC 5986: A large, loose globular cluster. With a six-inch (15-cm) or larger telescope, you should be able to resolve some of the brighter members. Magnitude: +7.5; distance: thirty-six thousand light-years.

LYNX (LINKS) | *The Lynx*

• **Abbreviation:** Lyn • **Genitive:** Lyncis • **Area:** 545 square degrees • **Size Ranking:** 28th

• **Best Viewed:** January to March • **Latitude:** 90ºN–28ºS • **Width:** • **Height:**

Lynx is composed of a chain of faint stars located immediately north of Gemini and Cancer. Introduced by Johannes Hevelius in 1690, the origins of the name are obscure. Some have suggested he named it the Lynx because it requires the celebrated sharp eyes of this animal to spot its form in the heavens.

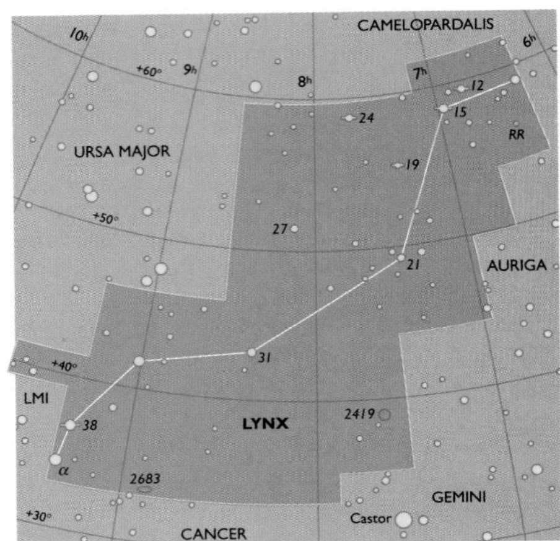

FEATURES OF INTEREST

12 LYNCIS: A triple-star system, with components A and B—magnitude +5.5 and +6—separated by 1.8 arc seconds. Component C, magnitude +7.5, is located 8.5 arc seconds distant.

19 LYNCIS: An easier triple to divide than 12 Lyncis. A and B magnitudes: +5.5 and +6.5; separation: 14.7 arc seconds. Component C: magnitude +11; 74.2 arc seconds distant.

NGC 2419: The Intergalactic Wanderer, so-called because it is believed to be independent of our Milky Way galaxy, unlike most globular clusters, which orbit the center of the galaxy. This is a very compressed globular cluster, visible with six-inch (15-cm) and larger instruments as a round small patch of light. More than two hundred thousand light-years distant—farther than the Magellanic Clouds—it is the most distant globular cluster visible with small telescopes.

NGC 2683: A spiral galaxy, viewed nearly edge on, with a conspicuous central bulge. Magnitude: +9.3, with a magnitude +13 star visible just off its northeastern tip.

LYRA (LYE-RAH) | *The Lyre*

• **Abbreviation:** Lyr • **Genitive:** Lyrae • **Area:** 286 square degrees • **Size Ranking:** 52nd

• **Best Viewed:** July to August • **Latitude:** 90ºN–42ºS • **Width:** • **Height:**

This small, but very conspicuous constellation was identified by the ancient Greeks as the musical instrument created from a tortoise shell by Hermes for Apollo, who later gave it to Orpheus. The constellation is notable for its blazing first-magnitude luminary, Vega, and a small number of other worthwhile objects for the amateur astronomer, including a classic quadruple star and the most famous planetary nebula in the sky, M57, the Ring Nebula.

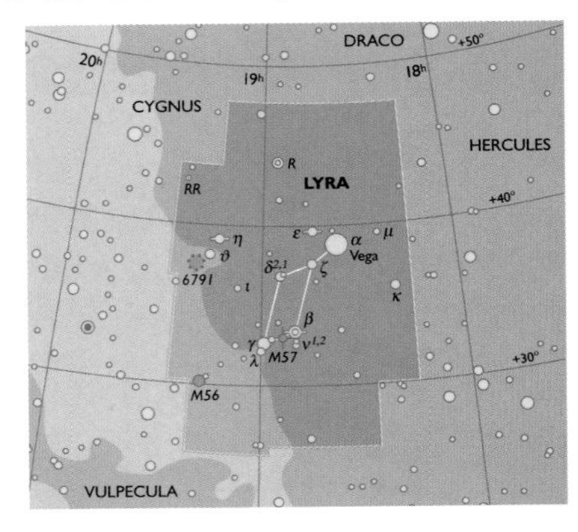

FEATURES OF INTEREST

ALPHA (α) LYRAE (VEGA): A brilliant white star, the fifth brightest in the sky at magnitude +0.03. This is the lead star in the prominent asterism known as the Summer Triangle. The other members are Deneb (Alpha α Cygni) and Altair (Alpha α Aquilae).

BETA (β) LYRAE: An eclipsing binary, with a brightness varying from magnitude +3.4 to +4.3 in a period of 12.94 days. Its variations can easily be followed by the novice by using the other stars in the constellation to judge its brightness.

DELTA 1 (δ1) AND DELTA 2 (δ2) LYRAE: Separated by 10.5 arc minutes, this pair may be split with the naked eye in a dark sky and is attractive in binoculars. Delta 1 is magnitude +5.51, Delta 2 is magnitude +4.52.

EPSILON 1 (ε1) AND EPSILON 2 (ε2) LYRAE: The renowned Double-Double quadruple star system is an interesting test for the small telescope. The two main components are easily seen in binoculars as an equal magnitude +5 pair, separated by almost four arc minutes. With a three-inch (8-cm) or larger telescope, the two stars can be split, with separations of 2.6- and 2.3-arc seconds.

MESSIER 56: A typical globular cluster, situated in a beautiful field of faint stars, that is rather loosely structured and ethereal-looking in a small telescope. It is only slightly brighter to the middle, magnitude +8.3, and resolution begins with eight-inch (20-cm) and larger telescopes. Distance: thirty-three light-years.

MESSIER 57: This is probably the easiest planetary nebula for a newcomer to astronomy to find in the night sky. It is located about halfway between Beta (β) and Gamma (γ) Lyrae. Its smoke-ring structure is visible at high magnification with a six-inch (15-cm) telescope, although the central star remains invisible to most amateur telescopes.

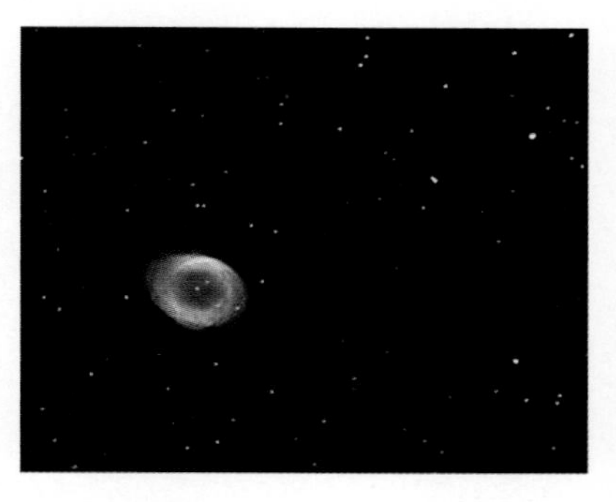

M57

This was the first nebula to be discovered (1779).

MENSA (MEN-SAH) | *The Table Mountain*

• **Abbreviation:** Men • **Genitive:** Mensae • **Area:** 153 square degrees • **Size Ranking:** 75th

• **Best Viewed:** December to February • **Latitude:** 5ºN–90ºS • **Width:** • **Height:**

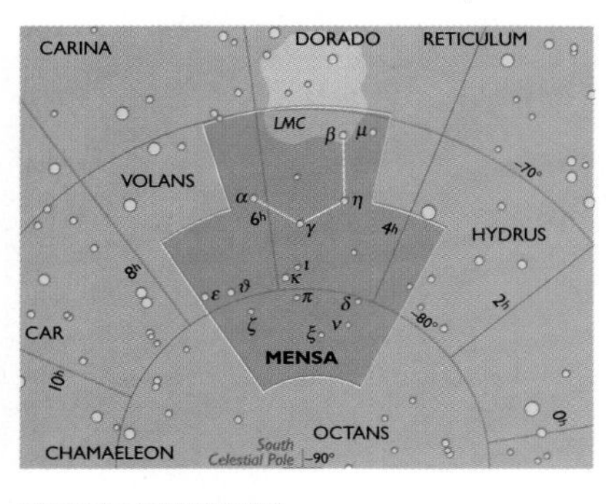

This constellation of the deep-southern sky was created by the French astronomer Lacaille in 1752. It is supposed to represent South Africa's Table Mountain, near where Lacaille did much of his work. The northern reaches of the extremely faint constellation are immersed in the Large Magellanic Cloud, which spills into it from the neighboring constellation Draco.

FEATURES OF INTEREST

ALPHA (α) MENSAE: A yellow dwarf star thirty light-years away; magnitude: 5.1.

GAMMA(γ) MENSAE: A wide optical double star, which requires at least a six-inch (15-cm) telescope to split in two. Component A: magnitude +5; component B: magnitude +11. The two are separated by 38.2 arc seconds.

MICROSCOPIUM (MY-KRO-SKO-PEE-UM) | *The Microscope*

• **Abbreviation:** Mic • **Genitive:** Microscopii • **Area:** 210 square degrees • **Size Ranking:** 66th

• **Best Viewed:** August to September • **Latitude:** 45ºN–90ºS • **Width:** • **Height:**

Another of the creations of Lacaille in 1752, this small constellation has few conspicuous stars and little to interest the small-telescope user. Its name honors the invention of the microscope in the late sixteenth century. Mid-northern observers with a clear southern horizon can see the double star Alpha Microscopii.

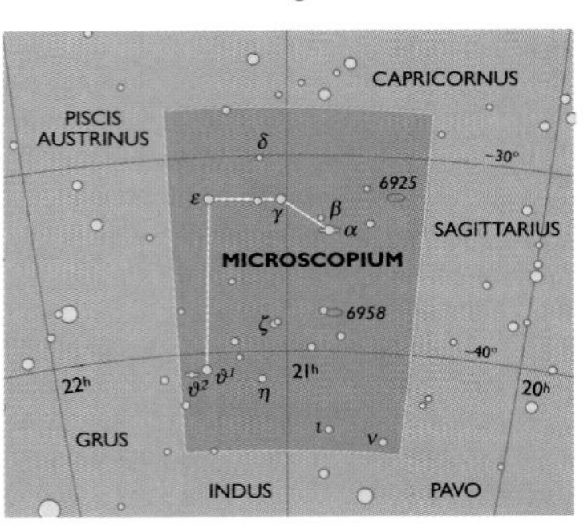

FEATURES OF INTEREST

ALPHA (α) MICROSCOPII: A wide double star, magnitudes +5 and +9.5, separated by 20.6 arc seconds.

NGC 6925: A spiral galaxy that is seen edge on, oriented due north/south. At magnitude 11.3, NGC 6925 is visible with six-inch (15-cm) and larger telescopes.

NGC 6958: Another spiral galaxy, a little smaller than NGC 6925, but of equal magnitude, with a very bright core.

MONOCEROS (MOH-NO-SER-US) | *The Unicorn*

• **Abbreviation:** Mon • **Genitive:** Monocerotis • **Area:** 482 square degrees • **Size Ranking:** 35th

• **Best Viewed:** January to February • **Latitude:** 78ºN–78ºS • **Width:** • **Height:**

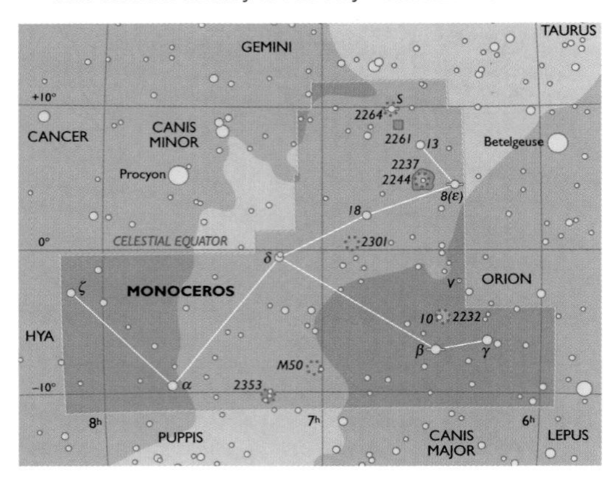

This constellation first appeared on a map drawn by the German astronomer Jakob Bartsch in 1624. Bartsch was the son-in-law of Johannes Kepler, who determined that the orbits of the planets are elliptical. Monoceros represents the mythical one-horned creature known as the unicorn. The constellation is located due east of Orion and is neatly bisected by the Milky Way. Consequently, Monoceros is well known for its profusion of star clusters and nebulae, as well as a great selection of multiple and variable stars. One Messier object is located here, the open cluster M50.

FEATURES OF INTEREST

BETA (β) MONOCEROTIS: An excellent triple star system for the small telescope. All three components are nearly equal in brightness: Component A is +4.6; component B, +5.22; and component C, +5.6. A and B are separated by 7.4 arc seconds; B and C by 2.8 arc seconds.

EPSILON (ε) MONOCEROTIS: A triple star, magnitudes +4.5, +6.5, and +12.5. The A and B components are separated by 13.2 arc seconds; the C component is located 93.7 arc seconds away.

MESSIER 50: This cluster is visible with binoculars and is an attractive sight with a small telescope: a roundish group of stars about ten arcminutes in diameter. Cluster members have a wide range of brightness—from magnitude +9 to +14. Up to one hundred stars are visible in the cluster.

NGC 2237/NGC 2244: This nebula and star cluster complex is known collectively as the Rosette Nebula. The wreath-shaped nebula is a rather elusive object for the small telescope, but eight-inch (20-cm) and larger instruments show something of the nebula as a smoky, grayish cloud surrounding a bright, coarse cluster of stars. The cluster it envelops, NGC 2244, is easily visible with binoculars and consists of about fifteen stars of magnitudes ranging from +6 to +9.

NGC 2264: This is a large, coarse cluster of bright and faint stars. Its most prominent member is S Monocerotis, a blue-white variable star. The outline of the cluster gives the group its proper name: the Christmas Tree cluster. The stars are surrounded by an area of faint nebulosity, including the dramatic-looking Cone Nebula.

NGC 2261: This is known as Hubble's Variable Nebula for its peculiar changes in brightness and shape as recorded on observatory photographs through the twentieth century. The light variations may be caused by moving clouds of dark, cold dust.

NGC 2301: A spectacular cluster of about fifty eighth magnitude and fainter stars oriented roughly north/south with a third spur of stars angling off toward the east.

NGC 2353: A very large, showy cluster of brilliant little jewels, arrayed to the northeast of a magnitude +6 star. At least forty stars are visible with a six-inch (15-cm) telescope.

NGC 2244
The stars of the open cluster NGC 2244 (at the center of this picture) are surrounded by the Rosette Nebula.

MUSCA (MUS-KAH) | *The Fly*

• **Abbreviation:** Mus • **Genitive:** Muscae • **Area:** 138 square degrees • **Size Ranking:** 77th

• **Best Viewed:** April to May • **Latitude:** 14ºN–90ºS • **Width:** • **Height:**

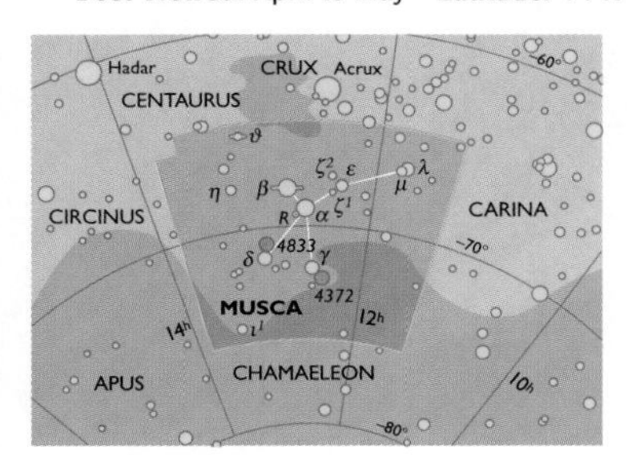

This far-southern constellation is a modern grouping, at first called Apis the Bee, before evolving into Musca the Fly around the time of Nicolas de Lacaille in the eighteenth century. Immersed in the southern Milky Way, this constellation is easy to find, lying immediately south of the Southern Cross.

FEATURES OF INTEREST

ALPHA (α) MUSCAE: A magnitude +2.7 star with a +12.8 magnitude companion 29.6 arc seconds away.

BETA (β) MUSCAE: A close double star, magnitudes +4, separated by 1.6 arc seconds.

NGC 4372: A large, slightly condensed globular cluster, with a core about five arc minutes across. Magnitude: +7.3; distance: sixteen thousand light-years.

NGC 4833: A fairly compact globular cluster. The brightest stars can be detected with a four-inch (10-cm) telescope. Magnitude: +7; distance: eighteen thousand light-years.

NORMA (NOR-MAH) | *The Carpenter's Level*

• **Abbreviation:** Nor • **Genitive:** Normae • **Area:** 165 square degrees • **Size Ranking:** 74th

• **Best Viewed:** June • **Latitude:** 29ºN–90ºS • **Width:** • **Height:**

Norma is the carpenter's level, a constellation created by Lacaille in 1752 and embedded in the Milky Way. Its brightest star is the fourth-magnitude Gamma (γ) Normae. Although there are a good number of clusters and nebulae within the confines of this small constellation, most of them are not particularly noteworthy.

FEATURES OF INTEREST

GAMMA 1 (γ1) AND GAMMA 2 (γ2) NORMAE: A double star divisible with binoculars. The primary, Gamma-2, is an orange giant; its companion, Gamma-1, is a yellow supergiant, more than forty times farther away than Gamma-2.

EPSILON (ε) NORMAE: A double star, magnitudes +4.5 and +7.5, separated by twenty-two arc seconds. The two are easily divisible with a three-inch (8-cm) or larger telescope.

NGC 6087: A scattered open cluster that lies in a rich Milky Way field. About forty stars of magnitudes +7 and fainter are visible.

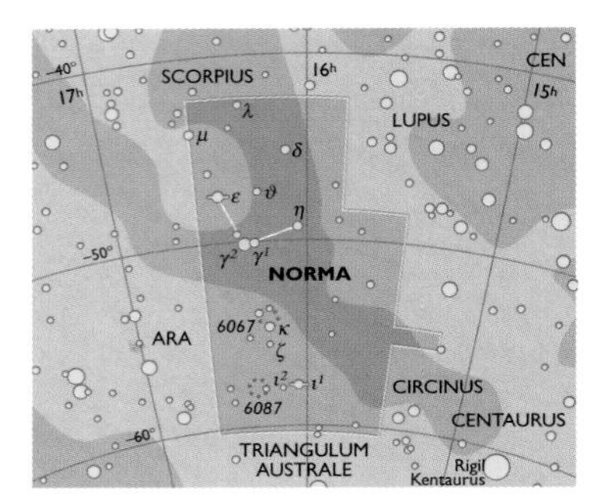

OCTANS (OCK-TANZ) | *The Octant*

• **Abbreviation:** Oct • **Genitive:** Octantis • **Area:** 291 square degrees • **Size Ranking:** 50th

• **Best Viewed:** October • **Latitude:** 0ºN–90ºS • **Width:** • **Height:**

This southern constellation lies in a rather undistinguished region of the sky that includes the south celestial pole. Its most notable feature is the faint star Sigma (σ) Octantis. The constellation was introduced in the year 1752 by Lacaille and is named in honor of the octant, a predecessor to the venerable navigational tool, the sextant.

FEATURES OF INTEREST

GAMMA 1 (γ1) GAMMA 2 (γ2) OCTANTIS: A double star divisible with binoculars.

LAMBDA (λ) OCTANTIS: A close double star, magnitudes +5.5 and +7.5, separated by 3.1 arc seconds.

SIGMA (σ) OCTANTIS: This unremarkable, magnitude +5.5 star has the singular distinction of marking the approximate position of the south celestial pole, although with time it is slowly moving away from that point. Currently, it is one degree away from true south.

R OCTANTIS: A variable star, ranging in magnitude from +6.6 to fainter than +13 over a period of 405 days.

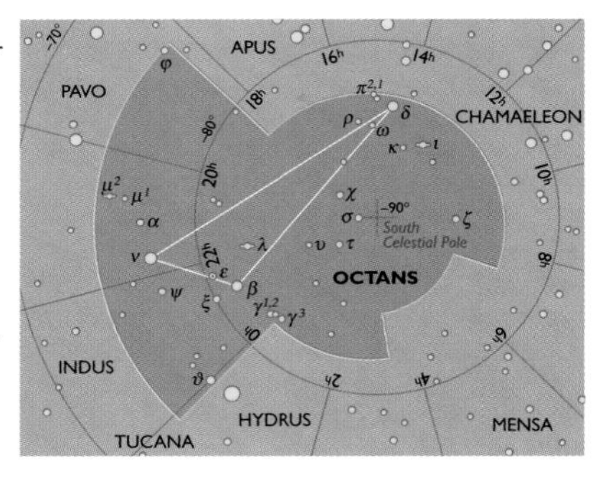

OPHIUCHUS (OF-FEE-U-KUS) | *The Serpent Bearer*

• **Abbreviation:** Oph • **Genitive:** Ophiuchi • **Area:** 948 square degrees • **Size Ranking:** 11th

• **Best Viewed:** June to July • **Latitude:** 59ºN–75ºS • **Width:** • **Height:**

In Greek mythology, Asclepius was known as the Serpent Bearer because of two snakes that he carried. The venom of one was deadly, but that of the other healed the sick. It was reputed that Asclepius could raise the dead. This angered Hades, god of the underworld, who asked Zeus to strike Asclepius dead. So doing, Zeus translated Asclepius' form to the stars. Partly immersed in the Milky Way, Ophiuchus is a good hunting ground for globular clusters, which tend to form a halo around the core of our galaxy. At least twenty can be detected with an eight-inch (20-cm) telescope, including seven Messier objects.

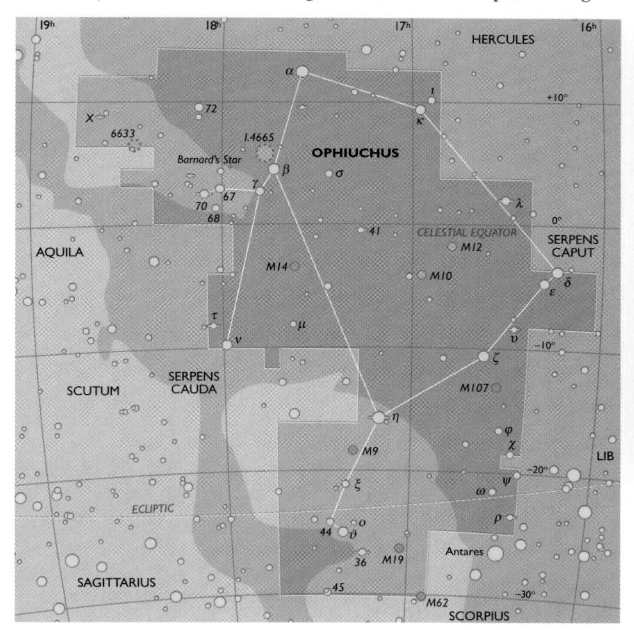

M10
With binoculars, this globular cluster appears half the size of the full moon.

FEATURES OF INTEREST

70 OPHIUCHI: A pair of stars that revolve around each other every eighty-eight years. The components are magnitude +4 and +6, separated by 3.7 arc seconds and widening.

BARNARD'S STAR: A faint red dwarf, the closest to the sun after the Alpha Centauri system (distance: 5.94 light-years). This star has the highest proper motion known: It is moving against the background of more distant stars by the diameter of the moon every 175 years. In 9,700 years, it will pass within 3.8 light-years of Earth.

MESSIER 9: Although M9 is a fairly loosely structured globular, it is small and its brightest stars are about fourteenth magnitude, requiring a ten-inch (25-cm) telescope to achieve some resolution. Magnitude +7.9.

MESSIER 10 AND MESSIER 12: Located 3.4 degrees apart near the center of the constellation, these two globular clusters are rather loosely structured and can be easily resolved with even a four-inch (10-cm) telescope under dark skies. Both clusters: magnitude +6.6.

MESSIER 14: This moderate-sized globular appears rather hazy in a small telescope. Telescopes ten inches (25 cm) and larger will resolve the brightest stars. Magnitude +7.5.

MESSIER 62: A bright, quite round globular cluster in a rich star field. Magnitude +6.6.

MESSIER 107: The faintest of the Messier globular clusters. Magnitude: +8.1.

NGC 6633 AND IC 4665: Both of these large scattered clusters are suitable for binoculars, which will resolve the brightest stars.

ORION (OH-RYE-UN) | *The Hunter*

• **Abbreviation:** Ori • **Genitive:** Orionis • **Area:** 594 square degrees • **Size Ranking:** 26th

• **Best Viewed:** December to January • **Latitude:** 79ºN–67ºS • **Width:** • **Height:**

This is, without question, the brightest, most easily recognizable constellation in the sky, with its three bright stars that define Orion's belt. Since the constellation straddles the equator, its entire form can be seen in all but the most extreme northern and southern locations. In Greek mythology, Orion was the son of Poseidon and Queen Euryale and became the greatest hunter among men. Orion's pride at his accomplishments got the best of him, however, and to punish him the gods sent a deadly scorpion to sting him. The goddess Artemis placed Orion and Scorpius in opposing hemispheres of the sky to keep them forever apart. Long-exposure photographs show the constellation awash in star-forming nebulae. The constellation contains the Messier objects M42, M43, and M78.

IC 434 (Horsehead Nebula)
The horse's head at center is roughly one light-year high. It can be seen only with very large amateur telescopes.

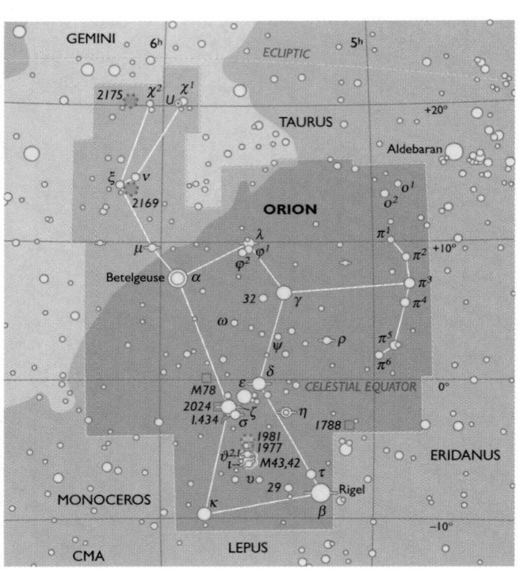

FEATURES OF INTEREST

ALPHA (α) ORIONIS (BETELGEUSE): This pulsating red giant ranges in magnitudes from +0.4 to +1.3 in irregular period, sometimes changing in brightness in just a few weeks. Betelgeuse (pronounced Bet-el-joose) is one of the largest stars known, more than six hundred times the diameter of our sun.

BETA (β) ORIONIS (RIGEL): A superluminous blue-white giant, as bright as fifty-five thousand suns. A companion star, magnitude +6.8, is nine arc seconds distant. The name Rigel comes from the Arabic word for foot and has been in use since the tenth century.

SIGMA (σ) ORIONIS: A beautiful five-star system, four of which are visible with a three-inch (8-cm) telescope. The main 3.8-magnitude star is flanked by two seventh-magnitude and one ninth-magnitude companions.

MESSIER 42: The most extraordinary of the Messier objects, visible to the naked eye and beautiful even with a three-inch (8-cm) telescope. This glowing cloud of dust, known as an emission nebula, is the birthplace of stars. In the heart of M42 lies Theta (ϑ) Orionis, the Trapezium: four bright stars that power the nebula. They are visible with the smallest of telescopes. The nebula itself is a complex region of light and shadow, looking like an enormous bird with outstretched wings. Across a dark gulf from the Trapezium is Messier 43, a seemingly detached portion of the nebula.

MESSIER 78: A large, diffuse reflection nebula that shines because of the light it reflects from nearby stars.

NGC 1788: A moderate-sized, diffuse reflection nebula.

NGC 2024: A large emission nebula with prominent dark zone, easily seen with an eight-inch (20-cm) telescope. It is located immediately east from Zeta Orionis. To see the nebula best, keep the bright star out of the field of view.

PAVO (PAH-VOH) | *The Peacock*

• **Abbreviation:** Pav • **Genitive:** Pavonis • **Area:** 378 square degrees • **Size Ranking:** 44th

• **Best Viewed:** July to September • **Latitude:** 15ºN–90ºS • **Width:** • **Height:**

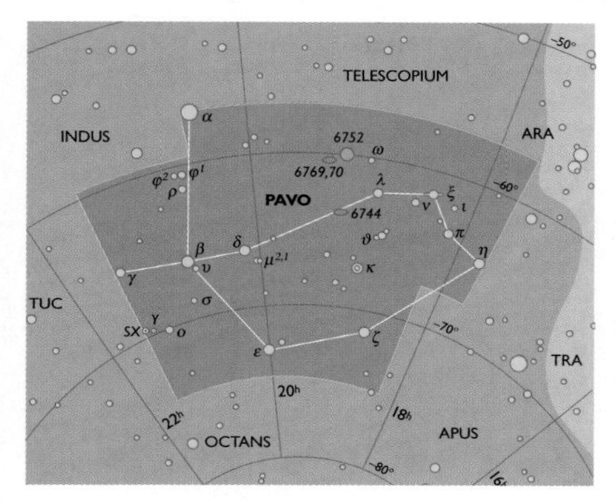

This constellation of the deep-southern skies was created by Johann Bayer in 1603 and represents a peacock, a name shared with its brightest star, Alpha (α) Pavonis. The name Pavo may trace its roots to Greek mythology and the story of the goddess Hera, who decorated the tail plumes of a peacock with the hundred eyes of Argus, a mythical monster. Located on the edge of the Milky Way between Telescopium and Octans, Pavo has one beautiful globular cluster and several galaxies.

FEATURES OF INTEREST

ALPHA (α) PAVONIS (THE PEACOCK STAR): This is the brightest star (+1.93) in and around Pavo, making it a good guide post. The star is actually part of a pair that orbit each other every twelve days. The two are too close to be split with an amateur telescope.

KAPPA (κ) PAVONIS: A Cepheid variable, which varies between magnitude +3.9 and +4.8 in a period of 9.065 days.

XI (ξ) PAVONIS: A double star with a 4.4 magnitude red giant paired with an eighth-magnitude companion.

NGC 6744: Usually classed as a barred spiral, this multiple-armed galaxy appears about one third the size that it does in the best photographs. Only the bright central region is likely to be detected with a small telescope, appearing as an elongated haze. Magnitude: +9.24.

NGC 6752: This is one of the finest globular clusters in the sky, not on a par with Omega Centauri or 47 Tucanae, but certainly a rival to Messier 13. It is fifteen arc minutes in diameter, with its brightest members around magnitude +11. Resolution of the outer edges is possible with even a three-inch (8-cm) telescope. Magnitude: +6; distance: seventeen thousand light-years.

NGC 6769 AND NGC 6770: These two galaxies are the brightest of a group of four, including NGC 6771 and IC 4842. High-contrast electronic images show a thin, clumpy extension of NGC 6770's spiral arm extending toward NGC 6769. None of this is visible, although the galaxies appear almost in contact. The two are influenced by each other's gravitational field. This is an interesting sight for ten-inch (25-cm) and larger telescopes.

NGC 6752
This large globular cluster is relatively close to Earth—seventeen thousand light-years distant.

PEGASUS (PEG-A-SUS) | *The Winged Horse*

• **Abbreviation:** Peg • **Genitive:** Pegasi • **Area:** 1,121 square degrees • **Size Ranking:** 7th

• **Best Viewed:** September to October • **Latitude:** 90ºN–53ºS • **Width:** • **Height:**

In Greek mythology, Pegasus sprang up from the blood of the severed head of Medusa, the Gorgon who was slain by Perseus. The winged steed immediately flew up into the heavens and later descended onto Mount Helicon, where he created the spring of Hippocrene, the source of inspiration for poets. The wild horse was eventually tamed by Athena and given to Bellerophon, who used Pegasus for his travels. On one of their journeys, Pegasus threw Bellerophon to his death and continued the voyage alone, eventually reaching heaven, where he was placed among the stars. The Great Square of Pegasus is the dominant asterism of mid-northern fall skies, and the whole area, far from the plane of the Milky Way, is strewn with galaxies both bright and faint. One Messier object can be found here, the globular cluster M15.

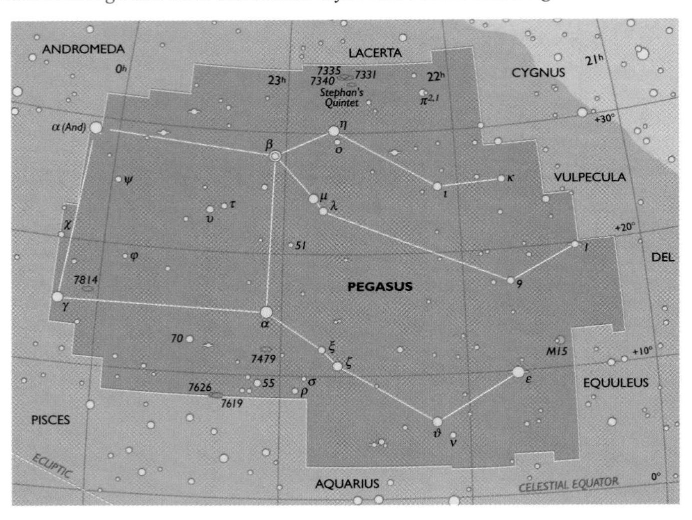

FEATURES OF INTEREST

BETA (β) PEGASI (SCHEAT): An irregular variable star with a magnitude that ranges from +2.2 to +2.74.

MESSIER 15: This is a fine, bright example of the globular star class, notable for its extremely compressed and blazing core. It can be seen with even the smallest telescope. An eight-inch (20-cm) instrument will help resolve the outer regions into hundreds of brilliant stars. Magnitude: +6.4; diameter: 6.8 arc minutes.

NGC 7331: A spiral galaxy that looks like a distant twin of M31, the Andromeda galaxy. At magnitude +9.5, NGC 7331 can be seen faintly with a three-inch (8-cm) instrument and is an interesting sight with an eight-inch (20-cm) telescope.

NGC 7331 *This is the brightest of Pegasus' dozen or so galaxies.*

NGC 7479: This is a good representative of Hubble's SBb-class of barred-spiral galaxy, although it is a little faint for small instruments. An eight-inch (20-cm) telescope will start to show the galaxy to advantage, as a straight bar with a brighter core.

NGC 7814: A fairly bright Sa- or Sb-class spiral galaxy seen edge on. It shows up well with an eight-inch (20-cm) telescope. Magnitude: +11.4.

STEPHAN'S QUINTET: Named after the French astronomer Edouard Stephan who discovered the group in the nineteenth century, Stephan's Quintet refers to the five most prominent members of a compact group of galaxies that has a sixth, fainter member (NGC 7320C). The galaxies are all about magnitude +14 and fainter, and one or more of the brighter members may be glimpsed with an eight-inch (20-cm) telescope.

PERSEUS (PURR-SEE-US) | *The Hero*

• **Abbreviation:** Per • **Genitive:** Persei • **Area:** 615 square degrees • **Size Ranking:** 24th

• **Best Viewed:** November to December • **Latitude:** 90°N–31°S • **Width:** • **Height:**

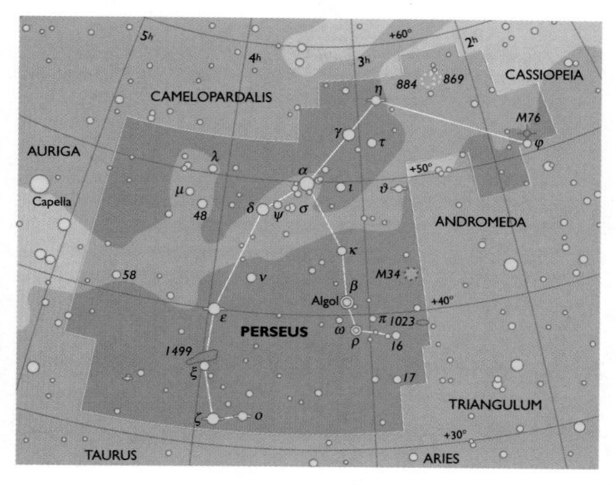

Perseus is famous in Greek mythology as the slayer of one of the Gorgons—a creature so ugly that a single look could turn a viewer to stone.

Perseus is the central character in a Greek myth involving many of the constellations visible in the mid-northern autumn sky. Slayer of the Gorgon Medusa, this hero used the head to kill the sea monster Cetus, turning the monster to stone. In the process Perseus saved Andromeda, who had been chained to the rocks in the monster's path. Straddling the autumn Milky Way, the constellation Perseus is a rich hunting ground for the amateur equipped with either binoculars or a telescope. This constellation features many bright stars and its outline is prominent even in suburban locales. The Milky Way is not as bright here as in nearby Cygnus, making it possible to see many bright star clusters, gaseous and planetary. The constellation includes two Messier objects, M34 and M76.

FEATURES OF INTEREST

BETA (β) PERSEI (ALGOL): This was the first recognized eclipsing variable, a star with a fainter companion star that regularly blocks the light of the primary. Algol's brightness varies between magnitudes +2.1 and +3.3 every 2.867 days.

NGC 869 AND 884: The famous Double Cluster—two of the heavens' best open star clusters. Visible to the naked eye as a bright, oblong nebulous patch in the Milky Way halfway between Perseus and Cassiopeia, this object is best seen with a small telescope equipped with a wide-field eyepiece. The clusters are not physically related: NGC 869 is seven thousand light-years away; NGC 884 is eleven hundred light-years farther.

NGC 1023: A lens-shaped elliptical galaxy that can show up well with four-inch (10-cm) and larger telescopes.

NGC 1499: A large, faint emission nebula illuminated by Zeta (ζ) Persei.

MESSIER 34: A rather coarse star cluster for the telescope, but an attractive sight with binoculars, which will resolve the brighter members. Magnitude: 5.2.

MESSIER 76: Located eight degrees southwest from the Double Cluster, M76 is one of only four planetary nebulae in Messier's famous catalog. Also known as the Little Dumbbell, the nebula measures approximately two arc minutes by one arc minute in size and has a rather high surface brightness. To be seen well, it will require at least a three-inch (8-cm) telescope magnifying about 120x.

PERSEIDS: This is the most famous of the annual meteor showers, with peak activity on August 11th and 12th. You can expect to see fifty to sixty meteors per hour. The shower was once known by the more poetic name The Tears of St. Lawrence, after the saint whose feast day is celebrated at this time of year.

PHOENIX (FEE-NIX) | *The Phoenix*

• **Abbreviation:** Phe • **Genitive:** Phoenicis • **Area:** 469 square degrees • **Size Ranking:** 37th
• **Best Viewed:** October to November • **Latitude:** 32ºN–90ºS • **Width:** • **Height:**

This is one of Johann Bayer's creations, delineated in the year 1603. Representing the bird of legend that rose from its own ashes, the small southern constellation is dominated by the magnitude +2.4 Alpha (α) Phoenicis. All the other stars are rather faint.

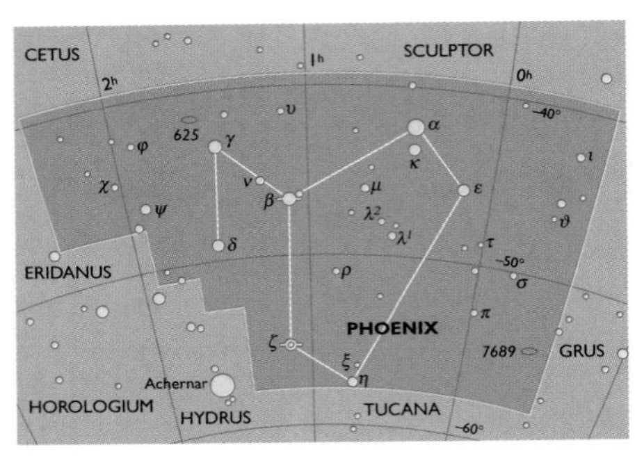

FEATURES OF INTEREST

BETA (β) PHOENICIS: A triple-star system, with magnitude +4 A and B components separated by 1.4 arc seconds. The third component, magnitude +11.5, can be found 57.5 arc seconds distant.

ZETA (ζ) PHOENICIS: A triple star, with A and B components that are too close to split with less than ten-inch (25-cm) telescopes. Component A is magnitude +4; component B is +7, separated by less than one arc second. Component C is magnitude +8 and 6.4 arc seconds distant. The brightest star is also an eclipsing binary, with a magnitude ranging from +3.9 to +4.4 in a period of 1.67 days.

NGC 625: This galaxy appears as an elongated oval of light, a little brighter along its central axis. Requires at least a six-inch (15-cm) telescope to be seen.

PICTOR (PIK-TOR) | *The Painter*

• **Abbreviation:** Pic • **Genitive:** Pictoris • **Area:** 247 square degrees • **Size Ranking:** 59th
• **Best Viewed:** December to February • **Latitude:** 26ºN–90ºS • **Width:** • **Height:**

Pictor the Painter is one of the constellations defined by Nicolas de Lacaille in the year 1752. The faint stars that make up the small southern constellation are difficult to trace, and there are very few objects of interest to the user of a small telescope or binoculars.

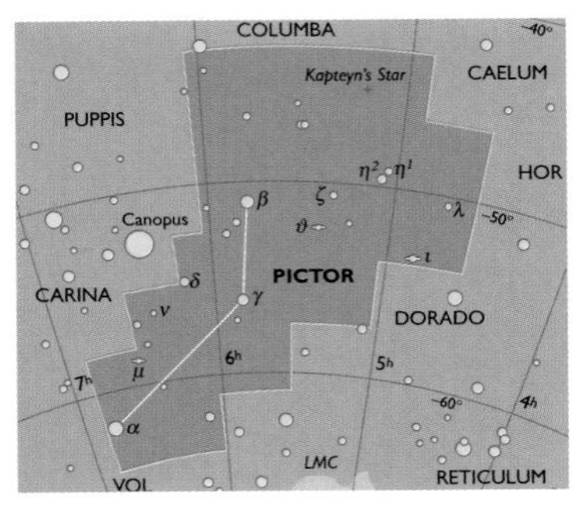

FEATURES OF INTEREST

BETA (β) PICTORIS: This nondescript fourth-magnitude star is the first star ever to display a protoplanetary disk of primordial material surrounding it—perhaps a future planetary system. This disk is invisible in amateur instruments.

IOTA (ι) PICTORIS: A double star with magnitudes +5.5 and +6.5, separated by 12.3 arc seconds.

KAPTEYN'S STAR: A red dwarf star of magnitude +8.8, 12.78 light years from the sun and displaying a large proper motion of 8.7 arc seconds per year—second only to Barnard's Star in Ophiuchus.

PISCES (PIE-SEEZ) | *The Fish*

• **Abbreviation:** Psc • **Genitive:** Piscium • **Area:** 889 square degrees • **Size Ranking:** 14th
• **Best Viewed:** October to November • **Latitude:** 83ºN–56ºS • **Width:** ✋✋✋ • **Height:** ✋✋

Pisces is one of a number of constellations in this region of the sky associated with water, including Cetus (the Whale), Pisces Austrinus (the Southern Fishes), and Aquarius (the Water Bearer). In ancient times, all cultures associated this star grouping with fish. Greek mythology relates that Pisces represents the aquatic forms of Heros and Aphrodite, who changed themselves into fish and plunged into the Euphrates River to escape the giant Typhon. The stars that make up the constellation are all faint, but from a viewing site with no lights the outline may be traced easily enough, especially the Circlet of Pisces, an asterism immediately south from Pegasus. The constellation contains one Messier object, the spiral galaxy M74.

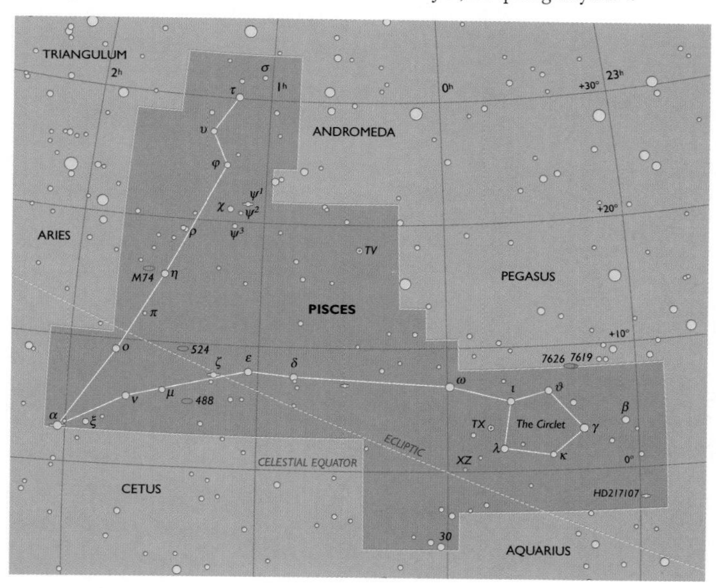

FEATURES OF INTEREST

ALPHA (α) PISCIUM: A binary star with an orbital period of 930 years. The components are magnitude +4.33 and +5.23, separated by 1.8 arc seconds.

PSI (Ψ) PISCIUM: A wide double star, both components magnitude +5, separated by thirty arc seconds. A third component, magnitude +11, lies 1.5 arc minutes away.

ZETA (ζ) PISCIUM: A double star easily split with a small telescope. Component A is magnitude +4.5, while component B is one magnitude fainter. The separation is 23.6 arc seconds. A third component, magnitude +12, orbits the B star and is only visible with large amateur telescopes.

TX PISCIUM: An irregular variable star, ranging in magnitude from +5.5 to +6.

MESSIER 74: Usually classed as one of the faintest Messier objects, M74 is a moderately large, face on Sc-type spiral galaxy. Telescopes six inches (15 cm) and larger reveal a bright core surrounded by a faint, circular haze. Magnitude: +9.2.

NGC 488: One of the brighter representatives of the many NGC galaxies in the region, this object appears as a slightly elongated glow oriented north/south with a bright core. Magnitude: +11.2.

NGC 524: A bright, E1-type elliptical galaxy, with well-defined edges and a bright core. Magnitude: +11.5.

NGC 7619 AND NGC 7626: This pair of elliptical galaxies, though rather faint for small telescopes, are of interest as the brightest members of the Pegasus I cloud of galaxies. With an estimated distance in excess of 150 million light-years, these galaxies are among the more remote visible with small telescopes. The pair are both about magnitude +12.

PISCIS AUSTRINUS (PIE-SIS OSS-TRIH-NUSS) | *The Southern Fish*

• **Abbreviation:** PsA • **Genitive:** Piscis Austrini • **Area:** 245 square degrees • **Size Ranking:** 60th

• **Best Viewed:** September to October • **Latitude:** 53ºN–90ºS • **Width:** ✋ • **Height:** ✋✋

This faint southern grouping is one of Ptolemy's original forty-eight constellations. For northern observers, it is marked by the first magnitude star Fomalhaut, the Solitary One, a beacon in a region noticeably bereft of bright objects.

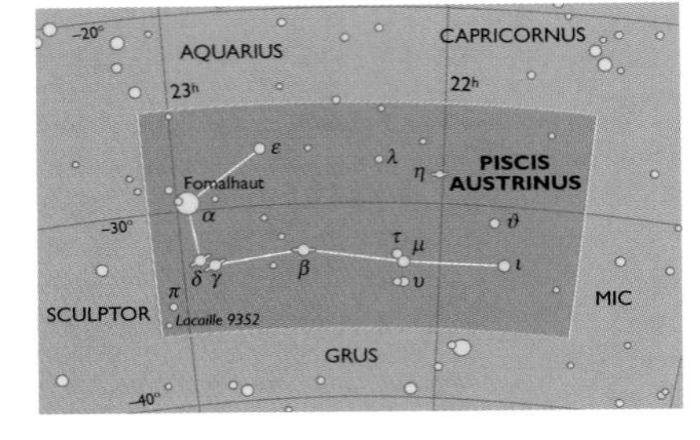

FEATURES OF INTEREST

ALPHA (α) PISCES AUSTRINI (FOMALHAUT): The name comes from the Arabic word for "mouth of the fish." Fomalhaut is one of the four Royal Stars of the Persians. Magnitude: +1.16.

BETA (β) PISCES AUSTRINI: A double star, magnitudes +4.5 and +7.5, with a wide separation of 30.4 arc seconds.

DELTA (δ) PISCES AUSTRINI: Another double-star system with magnitudes of +4.5, and +10. The two are separated by five arc seconds.

LACAILLE 9352: The fourteenth nearest star to the sun, this red dwarf exhibits the fourth highest proper motion in the heavens at a rate of 6.9 arc seconds per annum. Magnitude: +7.35; distance: 10.73 light years.

PUPPIS (PUP-IS) | *The Stern*

• **Abbreviation:** Pup • **Genitive:** Puppis • **Area:** 673 square degrees • **Size Ranking:** 20th

• **Best Viewed:** January to February • **Latitude:** 39ºN–90ºS • **Width:** ✋ • **Height:** ✋✋

Puppis is part of the ancient constellation Argo Navis, which was redrawn to include four separate constellations in 1877. A large, moderately bright grouping of stars, it is a wonderful area of the sky to sweep with a pair of binoculars, owing to its position along the Milky Way and the large number of bright open star clusters that populate this region of the sky.

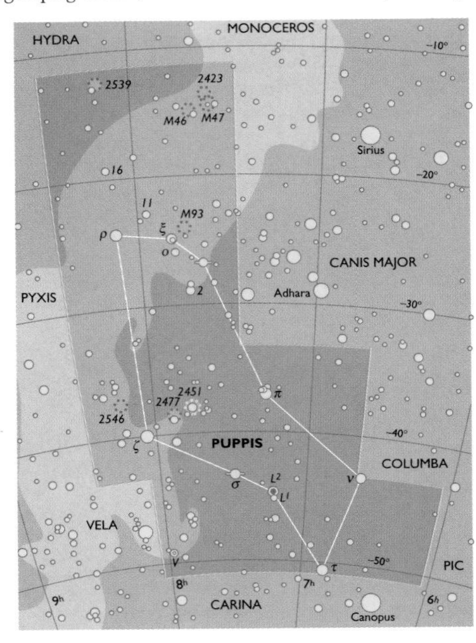

FEATURES OF INTEREST

ZETA (ζ) PUPPIS (NAOS): One of the largest blue supergiant stars in the Milky Way. Its magnitude +2.25 may not seem that bright, but it is fourteen hundred light-years away. If it were as close as stars such as Alpha Centauri or Sirius, it would dominate the night sky. The name Naos comes from the Greek word for shipping.

V PUPPIS: An eclipsing variable star; magnitude of +4.74 to +5.25 in 1.45-day period.

MESSIER 46: Fine, rich cluster of about 150 stars, most hovering around magnitude +10. A planetary nebula, NGC 2438, is visible in front to the northeast with four-inch (10-cm) and larger telescopes. Cluster size: about twenty-five arc minutes; magnitude +6.1.

MESSIER 47: This is a coarse grouping of bright and faint stars, not as rich as M 46 but brighter, with a total magnitude of +4.4. Shows up well in binoculars.

MESSIER 93: A bright, rich open star cluster, compressed toward the center and about twenty arc minutes across. Magnitude: +6.2.

NGC 2423: A fairly bright cluster about ten arc minutes in diameter. The stars, primarily magnitude +11, are grouped around a magnitude +9 luminary.

NGC 2451: A large cluster featuring many bright stars; well seen with binoculars.

NGC 2477: An extremely rich cluster, similar in structure to Messier 11 in Scutum. Includes about three hundred stars, predominantly magnitude +11, compressed toward the center. Total magnitude: about +7.

NGC 2539: A superb, well-resolved cluster, located just west of the bright field star 19 Puppis. There are about eighty cluster members, magnitudes +11 to +12, in a twenty-arc-minute area, with many subgroups of from three to ten stars.

PYXIS (PIK-sus) | *The Compass*

• **Abbreviation:** Pxy • **Genitive:** Pyxidis • **Area:** 221 square degrees • **Size Ranking:** 65th

• **Best Viewed:** February to March • **Latitude:** 52ºN–90ºS • **Width:** • **Height:**

The stars of this southern constellation were once included in a large grouping depicting Argo, the ship used by Jason and the Argonauts. But the French astronomer Nicolas de Lacaille saw fit to designate these stars as a separate group in 1752. The constellation represents a compass and is a rather small grouping of faint stars.

FEATURES OF INTEREST

EPSILON (ε) PYXIDIS: A binary system, with stars of magnitudes +5.5 and +9.5, separated by 17.8 arc seconds.

T PYXIDIS: A recurrent nova, usually about magnitude +14, but during an outburst it can attain magnitude +7. The outbursts can last for a hundred days or more. The last one was in 1966.

RETICULUM (REH-TIK-U-LUM) | *The Reticle*

• **Abbreviation:** Ret • **Genitive:** Reticuli • **Area:** 114 square degrees • **Size Ranking:** 82nd

• **Best Viewed:** December • **Latitude:** 23ºN–90ºS • **Width:** • **Height:**

Reticulum is another of Lacaille's creations, and commemorates the network of fine markings in a measuring eyepiece, known as a reticle. A small, far-southern constellation, the area features a few interesting double stars and relatively bright galaxies.

FEATURES OF INTEREST

ZETA 1 (ζ1) AND ZETA 2 (ζ2) RETICULI: A double star easily divisible with binoculars. Magnitudes +5 and +5.5, the stars are separated by 130 arc seconds. Both are similar in composition to our sun.

THETA (ϑ) RETICULI: A magnitude +6 and +8 double-star system, separated by 4.1 arc seconds.

NGC 1313: A bright spiral galaxy, with a bright north/south bar and faint mottled spiral arms. Magnitude: +9.4.

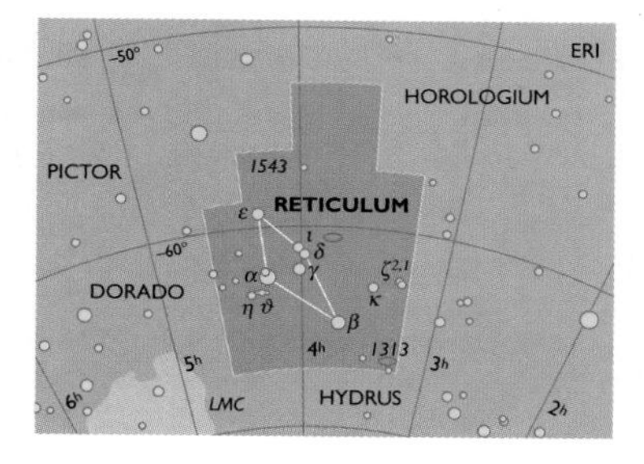

SAGITTA (SA-JIT-AH) | *The Arrow*

• **Abbreviation:** Sge • **Genitive:** Sagittae • **Area:** 80 square degrees • **Size Ranking:** 86th

• **Best Viewed:** August • **Latitude:** 90ºN–69ºS • **Width:** • **Height:**

In Greek mythology, Sagitta was the arrow used by Apollo to defeat the Cyclops. The Persians, too, recognized an arrow in this group of stars. Immersed in the Milky Way, the area is a fine sweep for binoculars.

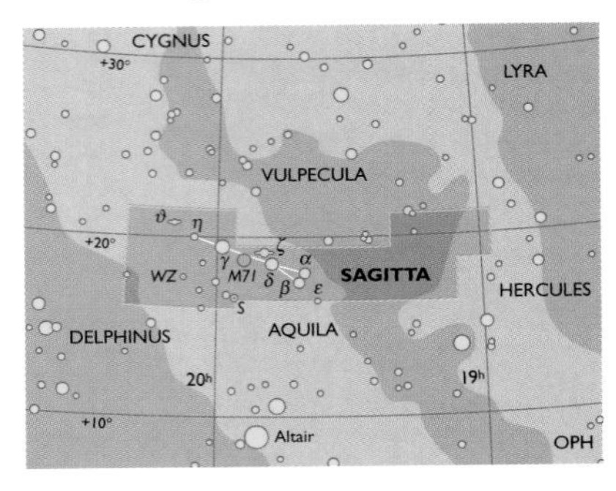

FEATURES OF INTEREST

EPSILON (ε) SAGITTAE: A double star that can be split with binoculars. The two stars are magnitudes +5.5 and +7.5, separated by 89.2 arc seconds.

WZ SAGITTAE: A recurrent nova, fainter than magnitude +16 at minimum and therefore invisible with all but the largest amateur telescopes. Typically, it rises very suddenly to a maximum of about +7 before fading to normal in a period of about two months. These outbursts sometimes occur decades apart.

MESSIER 71: For some time, astronomers were not sure whether to classify this object as an open or globular cluster. Now known to be a globular, M71 is a hazy, ethereal object with small telescopes. Apertures larger than eight inches (20 cm) permit resolution of the stars in the cluster. Magnitude: +8.3.

SAGITTARIUS (SADGE-IH-TARE-EE-US) | *The Archer*

• **Abbreviation:** Sgr • **Genitive:** Sagittarii • **Area:** 867 square degrees • **Size Ranking:** 14th

• **Best Viewed:** July to August • **Latitude:** 44ºN–90ºS • **Width:** • **Height:**

Lying in front of the richest region of the Milky Way, this is an extraordinary constellation to explore with binoculars and an almost limitless treasure trove for the small-telescope user. The center of our galaxy lies here, but is blocked by the star clouds and dust lanes of the Sagittarius Arm of the Milky Way. Mythologically, Sagittarius represents the fantastic creature known as the Centaur—half man, half horse. In ancient star atlases, Sagittarius was always portrayed in an aggressive, warlike pose with bow drawn, ready to pierce the heart of the Scorpion that precedes it in the heavens. Sagittarius was said to be placed in the heavens as a guide for the Argonauts on their quest for the Golden Fleece. The majority of the deep-sky objects here are all members of our home galaxy, including emission and dark nebulae, open and globular star clusters, and planetary nebulae. Sagittarius is home to no fewer than fifteen Messier objects—the most of any one constellation. Many of them can be easily seen with binoculars.

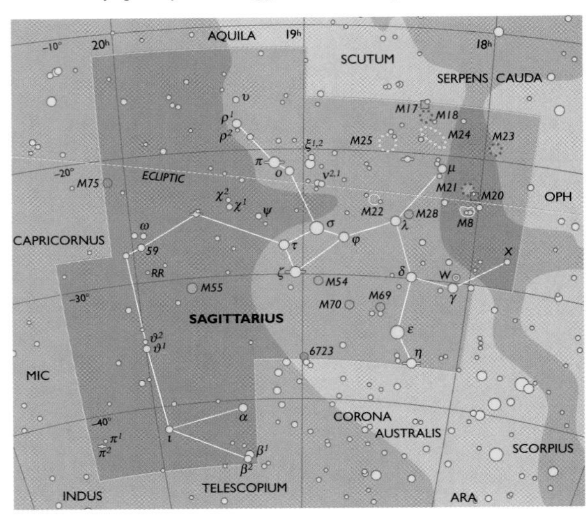

FEATURES OF INTEREST

MESSIER 8: This bright emission nebula is better known as the Lagoon Nebula, so-called because of a dark channel of obscuring matter that bisects this cloud of glowing gas. Easily seen with binoculars, the nebula envelops the coarse, open cluster NGC 6530.

MESSIER 17: One of the brightest emission nebulae in the sky, M17 is also known as the Swan or the Omega Nebula. The brightest portion is reminiscent of the body of a swan floating on a lake. A nebula filter attached to an eight-inch (20-cm) or larger telescope will bring out the fainter nebulosity, which almost fills the field. This is an easy object to detect with binoculars.

MESSIER 18: A small, coarse cluster of about a dozen magnitude +9 to +10 stars.

MESSIER 20: Also known as the Trifid Nebula, this combination reflection and emission nebula is beautiful in color photographs, which bring out the pinkish and bright blue portions of the nebula. For the small-telescope user, it is a more subtle object. Dark channels of obscuring matter divide the nebula into three distinct portions, hence its name. A prominent triple star lies in the brightest, central portion of the nebula.

MESSIER 21: A bright, rather small and coarse cluster of stars, readily seen with a small telescope. Total magnitude: +5.9.

MESSIER 22: One of the finest globular star clusters in the heavens, outranked only by the far southern clusters Omega Centauri and 47 Tucanae. About eighteen arc minutes in diameter, M22 is a rather loosely structured globular cluster, very well resolved with an eight-inch (20-cm) telescope, which will show many individual members even in the core of the cluster. M22 is one of the nearer globular clusters at a distance of about ten thousand light-years. Total magnitude: +5.1.

M8

The cluster of stars NGC 6530 is enveloped by the Lagoon Nebula, which covers twice the area of the full Moon in the night sky.

MESSIER 23: A large and attractive open star cluster of around a hundred magnitude +9 and fainter stars. Magnitude: +5.5.

MESSIER 24: Also known as the Small Sagittarius Star Cloud, this object is more than one degree in diameter and not a true cluster, but rather a brighter portion of the Milky Way. It can be seen with the naked eye, but is more impressive with binoculars. A small telescope will show field after field of stars beyond counting.

MESSIER 25: A bright, coarse, and scattered cluster of about fifty stars of magnitudes +6, plus fainter stars.

MESSIER 28: A small, moderately bright, but compressed globular star cluster, needing about a twelve-inch (30-cm) aperture to resolve the individual stars. Magnitude: +6.9.

MESSIER 54: A moderately bright, but small and very compressed globular cluster requiring a fairly large telescope to resolve. Magnitude: +7.7; distance: about sixty-eight thousand light-years.

MESSIER 55: A bright and very loosely structured globular cluster, easy to resolve with an eight-inch (20-cm) aperture. Magnitude: +7.0; distance: about twenty thousand light-years.

MESSIER 69: A small and quite round globular cluster, fairly compressed, with some resolution of fringe stars possible with an eight-inch (20 cm) aperture at high magnification. Magnitude: +7.7.

M20

More than five thousand light-years from Earth, M20 contains hot young stars surrounded by dust and gas. It is also known as the Trifid Nebula.

SCORPIUS (SKOR-PEE-USS) | *The Scorpion*

• **Abbreviation:** Sco • **Genitive:** Scorpii • **Area:** 497 square degrees • **Size Ranking:** 33rd

• **Best Viewed:** June to July • **Latitude:** 44ºN–90ºS • **Width:** • **Height:**

In Greek mythology, the scorpion is the creature that caused the demise of the great hunter Orion. It has also been associated with the legend of the boy Phaeton, who brashly attempted to drive the Chariot of the Sun across the heavens. The scorpion appeared and caused panic among the horses, and Phaeton was thrown to his death. One of the zodiac constellations, Scorpius is dominated by the supergiant star Antares and is almost totally immersed in the Milky Way. Unfortunately for mid-northern latitude observers, many of its greatest treasures are a little too far south to be seen. The constellation contains four Messier objects: M4, M6, M7, and M80.

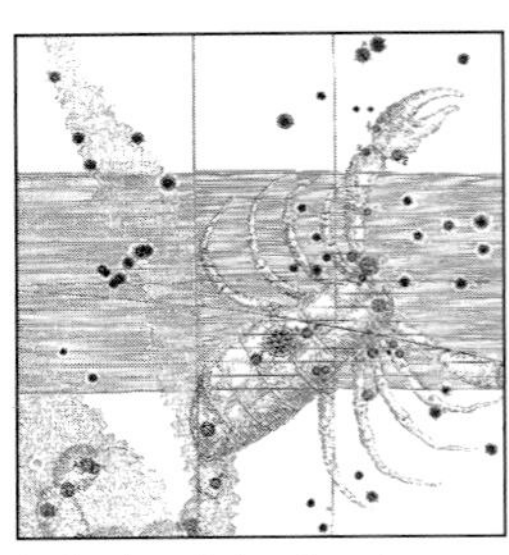

In Greek mythology, Scorpius killed Orion with its sting.

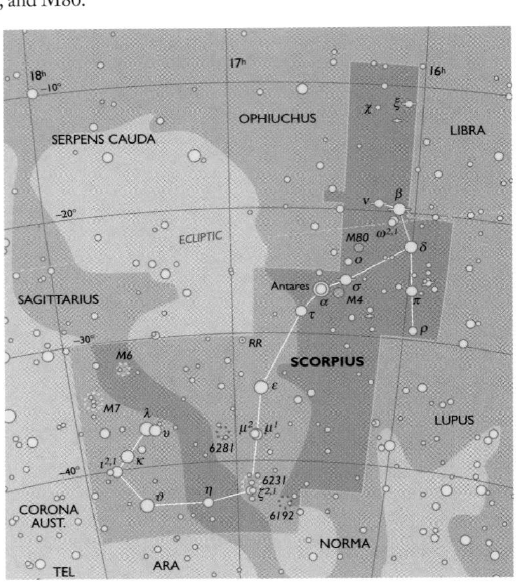

FEATURES OF INTEREST

ALPHA (α) SCORPII (ANTARES): This is a red supergiant star, one of the largest known, with a diameter more than six hundred million miles (966,000,000 km). If placed at the center of our solar system, the first four planets and a good portion of the asteroid belt would orbit inside it. Antares is also a double star with a faint companion, magnitude +5.37,three arc seconds west of the primary. Antares was one of the four Royal Stars of the Persians, guardians of heaven. Magnitude: +0.9.

NU (ν) SCORPII: A quadruple star, although A and B components, with magnitudes of +4.1 and 6.2 and a separation of only 1.3 seconds, are not easy to resolve. C and D components, forty-one arc seconds distant, are magnitudes +6.8 and +7.5, separated by 2.3 arc seconds.

XI (ξ) SCORPII: A triple-star system, with A and B components orbiting each other every 45.7 years. These stars were closest together in 1996 and cannot be split, although resolution may be possible in the early twenty-first century with medium-size telescopes. The third star, magnitude +7, was 7.4 arc seconds distant in 1959.

MESSIER 4: A large, bright, and loosely structured globular cluster with a prominent double row of bright stars crossing the core. Resolution is easy with a six-inch (15-cm) telescope. Magnitude: 5.9; distance; seven thousand light-years.

MESSIER 6 AND MESSIER 7: These brilliant clusters, located northeast of the scorpion's stinger, are easily seen together with binoculars, which will resolve both clusters. Each contains about fifty members, although Messier 7 is about twice the size of Messier 6. Both clusters are very young—less than seventy million years of age.

MESSIER 80: A compact, bright globular cluster that resists resolution into individual stars with less than a six-inch (15-cm) telescope. Magnitude: +7.2.

NGC 6192: A rather small but rich open cluster of about seventy-five stars, magnitudes +11 to +14.

NGC 6231: A superb, rich star cluster with a large range in brightness of members, from magnitudes +7 to +13. More than a hundred stars are visible in a fifteen-arc-minute field.

NGC 6281: A small cluster of about twenty-five stars, +9 to +11.

M4
This globular cluster is one of the closest of its kind to Earth. It can easily be seen with a small telescope.

SCULPTOR (SKULP-TOR) | *The Sculptor*

• **Abbreviation:** Scl • **Genitive:** Sculptoris • **Area:** 475 square degrees • **Size Ranking:** 36th
• **Best Viewed:** October to November • **Latitude:** 50ºN–90ºS • **Width:** ✋✋ • **Height:** ✋

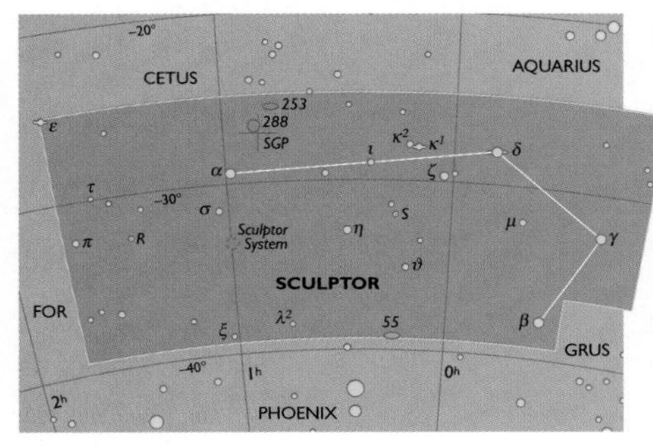

Originally known as the Sculptor's Workshop, this faint southern constellation was a creation of Nicolas de Lacaille in 1752. It is located ninety degrees from the plane of the Milky Way. The chief items of interest here are several members of a nearby galaxy group and one of the Milky Way's dwarf elliptical companions, known as the Sculptor System, members of which include NGC 55 and NGC 253.

FEATURES OF INTEREST

EPSILON (ε) SCULPTURIS: A double star featuring magnitude +5.5 and +9.5 components separated by 4.7 arc seconds.

R SCULPTURIS: A bright, red, variable star with a magnitude range from +6.1 to +8.8.

NGC 55: Classed as an S-B-type spiral, this large, ethereal object is oriented edge on to our line of sight. This is a fairly "dusty" system, evidence of which can be seen in twelve-inch (30-cm) and larger telescopes. It is a member of the Sculptor Group of galaxies, a scattered collection of galaxies that form the closest group of galaxies to our own Local Group. Magnitude: +8.22; size: twenty-five by three arc minutes; distance: 10.1 million light-years.

NGC 253: This is the brightest member of the Sculptor Group, which includes the galaxies NGC 55, NGC 247, NGC 300, and NGC 7793. Classed as an Sc-type spiral, the galaxy is oriented almost edge on; its structure is more nearly "classic" compared to NGC 55, which appears somewhat irregular in form. In binoculars, the galaxy looks like a thick streak. Photographs show a great deal of dust in the spiral arms, although this is beyond the reach of a small telescope. The core is fairly bright and elongated along the galaxy's major axis. It measures twenty-two by six arc minutes. Magnitude: +7.1; distance: 13.7 million light-years.

NGC 288: A moderately bright, loosely structured globular cluster. With six-inch (15-cm) and larger telescopes, individual stars can be resolved.

SCUTUM (SKU-TUM) | *The Shield*

• **Abbreviation:** Sct • **Genitive:** Scuti • **Area:** 109 square degrees • **Size Ranking:** 84th

• **Best Viewed:** July to August • **Latitude:** 74ºN–90ºS • **Width:** ✋ • **Height:** ✋

Originally known as Scutum Sobiescianum, this constellation represents the shield of the seventeenth century Polish king, John Sobieski, honored by Hevelius in his renowned 1690 star atlas. The component stars are faint and hard to locate since they are found within the Scutum Star Cloud, a bright patch in the Milky Way. Two Messier objects are located in Scutum, the open clusters M26 and M11.

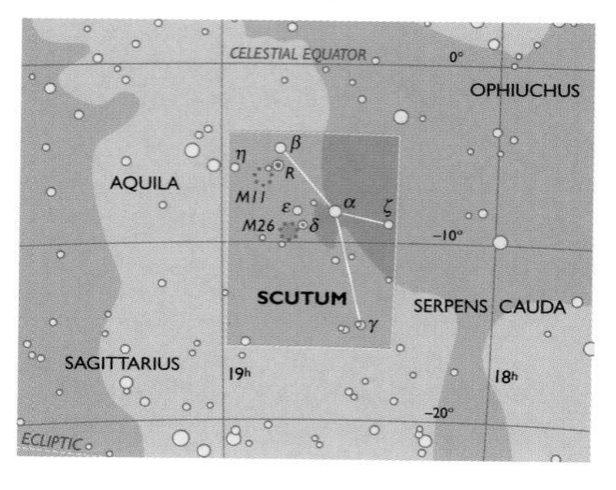

FEATURES OF INTEREST

MESSIER 11: Also known as the "Wild Duck," M11 is one of the richest open clusters in the sky and a beautiful sight with any telescope. Small and extremely compact, the cluster is wedge-shaped with component stars very similar in brightness, except for two magnitude +8 and +9 stars. Magnitude: +5.8.

MESSIER 26: A coarser group of stars, notable for its unusual structure. The southeastern part of the cluster is dominated by a diamond-shaped asterism of magnitude +9 and +10 stars, the interior of which is virtually starless. A chain of fainter stars extends northwest before curving north. Magnitude: +8.

SERPENS CAPUT & CAUDA (SIR-PENZ KAH-PUT / COW-DUH) | *The Snake's Head & Tail*

• **Abbreviation:** Ser • **Genitive:** Serpentis • **Area:** Caput 429, Cauda 208 square degrees

• **Size Ranking:** 23rd • **Best Viewed:** June to August • **Latitude:** 74ºN–64ºS • **Width:** ✋✋✋ • **Height:** ✋✋

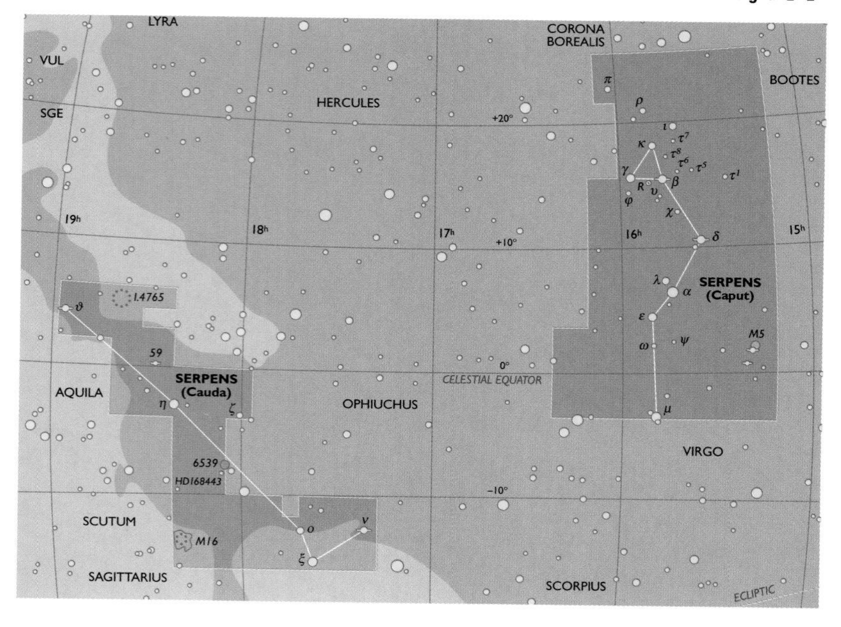

This constellation is divided into two parts, separated by Ophiuchus. The eastern half, Serpens Cauda, borders some of the richest star clouds of the Milky Way, while the western half, Serpens Caput, is well away from the plane of the Milky Way. Consequently, there are a wide variety of deep-sky objects for the small telescope. Two excellent Messier objects are located here, the globular cluster M5 and the emission nebula/open cluster complex M16.

M16

Taken by the Hubble Space Telescope, this photo of M16 (the Eagle Nebula) shows columns of dust against the backdrop of clouds of glowing gas.

FEATURES OF INTEREST

NU (ν) SERPENTIS: A wide double star, magnitudes +4.5 and +8.5. The two are separated by forty-six arc seconds.

THETA (ϑ) SERPENTIS: Another wide double, divisible with a small telescope. The components are magnitudes +4.59 and +4.99, separated by twenty-two arc seconds.

MESSIER 5: In the northern half of the sky, perhaps only M13 is a finer specimen of a globular cluster. This is a very bright, loosely compressed cluster with an enormous, assymetrical and easily resolved halo of stars surrounding a bright core. Magnitude: +5.8; distance: twenty-six thousand light-years. The bright double star 5 Serpentis, magnitudes +5 and +10, is in the field to the southeast.

MESSIER 16/NGC 6611: This combination nebula and star cluster is an intriguing object with an eight-inch (25-cm) telescope, appearing like a cosmic mushroom, the stem of which is formed by the star cluster and the head by the emission nebula. Small scopes (under four-inch [10-cm] aperture) are likely to show only the cluster, which has about sixty magnitude +8 and fainter stars.

IC 4756: With a diameter in excess of one degree, this coarse cluster of bright and faint stars is best seen with binoculars, which will reveal many of the brightest stars. IC 4756 includes about eighty stars of magnitude +7 and fainter.

SEXTANS (SEX-TANZ) | *The Sextant*

• **Abbreviation:** Sex • **Genitive:** Sextantis • **Area:** 314 square degrees • **Size Ranking:** 47th

• **Best Viewed:** March to April • **Latitude:** 78ºN–83ºS • **Width:** • **Height:**

A rather faint group of stars due south of Regulus, this constellation was created by Hevelius in 1680 to honor the instrument he used to measure the position of stars. There are a handful of galaxies here—bright enough for a small telescope—as well as a few interesting stars.

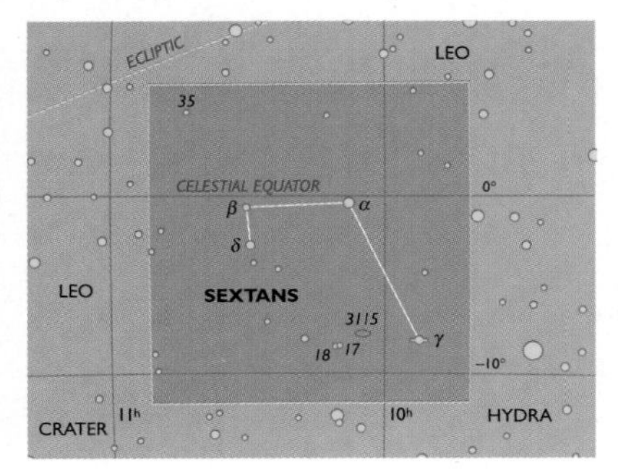

FEATURES OF INTEREST

17 AND 18 SEXTANTIS: A pair of sixth-magnitude stars divisible with binoculars.

35 SEXTANTIS: An attractive double star for the small telescope with components of magnitudes +6.5 and +7.5. The separation is 6.4 arc seconds.

NGC 3115: A bright spindle galaxy is lenticular (50) galaxy, suitable for a small telescope, featuring a bright core and tapered extremities. The galaxy is large, measuring 8.3 by 3.2 arc minutes, and has a magnitude of +9.2.

TAURUS (TORR-US) | *The Bull*

• **Abbreviation:** Tau • **Genitive:** Tauri • **Area:** 797 square degrees • **Size Ranking:** 17th
• **Best Viewed:** December to January • **Latitude:** 88ºN–58ºS • **Width:** • **Height:**

Recognized as representing the head of a bull since ancient times, this constellation assumed an important role in many of the myths of early civilization. Greek myth equated the constellation with the Minotaur, the half-man, half-bull monster. Later Greek legend relates the story of a snow-white bull that set itself among the herds of the Phoenician king's daughter, Europa. The bull was actually Zeus in disguise, madly in love with the king's lovely daughter. Gaining her trust, he carried her off to the island of Crete, where he once again assumed his divine form. Taurus is a dominant constellation of the northern skies, and a harbinger of winter. It is a beautiful region to sweep with binoculars. There are also two Messier objects located here, M1 and M45.

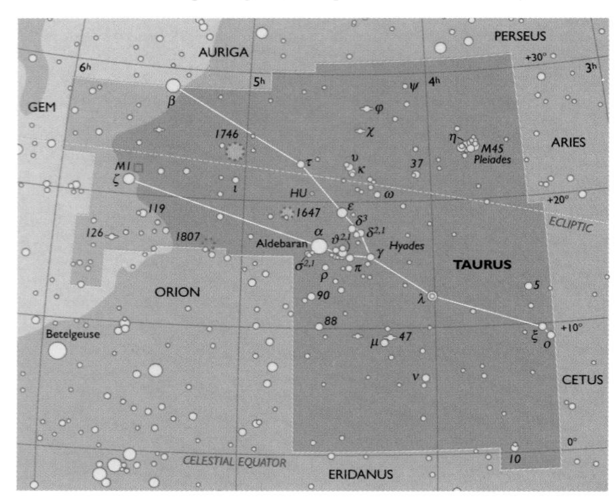

FEATURES OF INTEREST

ALPHA (α) TAURI (ALDEBARAN): One of the four Royal Stars of the Persians, which in 3,000 B.C. marked the solstice and the equinox points in the heavens. This brilliant star represents the eye of the bull and draws one's gaze to the Hyades, the V-shaped form of which delineates the head of the bull. Magnitude: +0.85.

THETA 1 (ϑ1) AND THETA 2 (ϑ2): With a separation of 3.5 arc minutes, this pair of stars in the Hyades star cluster, magnitudes +3.5 and +4, can be seen with the naked eye. Theta 2 is the brighter star.

LAMBDA (λ) TAURI: An eclipsing binary, with a magnitude that ranges from +3.4 to +4.1 every 3.95 days.

THE HYADES: This is one of the closest open clusters to our solar system, easily seen with the naked eye. Astronomers have used this cluster to calibrate the distance scale to other stars in our galaxy. The Hyades is best viewed with binoculars and has a distinctive V-shape. Aldebaran marks the tip of the left portion of the V, but is not associated with the cluster. There are more than 130 stars brighter than magnitude +9 in the group.

MESSIER 1 (THE CRAB NEBULA): This object is the remnant of a supernova, the only example of this class of object in the Messier Catalog. Associated with the appearance of a "guest star" in ancient Chinese records, it burst into view on July 4th, A.D. 1054. With a three-inch (8-cm) telescope, the nebula appears as an irregularly shaped, milky glow. The wispy extensions revealed in photographs can be seen only with very large amateur telescopes. Magnitude: +8.4.

The brightest of the stars at lower center right is Aldebaran, one of the eyes of Taurus. Aldebaran is a red giant located sixty light-years from Earth.

M45
Also known as the Pleiades, or the Seven Sisters, M45 is a cluster of young stars four hundred light-years from Earth.

MESSIER 45 (THE PLEIADES): This the best-known open cluster in the heavens, easily visible to the naked eye. Most people are able to see at least six tightly grouped stars under a dark sky. Binoculars reveal a beautiful sight with anywhere from thirty-five to fifty stars visible under ideal conditions. Telescopes bigger than ten inches (25 cm) easily reveal the associated reflection nebulae.

NGC 1647: A loose, scattered cluster of about twenty-five magnitude +8 and fainter stars.
NGC 1746: About fifty stars, magnitude +8 and fainter, visible in a forty-five-arc minute area. The eastern portion is separately cataloged as NGC 1758.
NGC 1807: A small, weak cluster of about fifteen magnitude +8 and +9 stars.

TELESCOPIUM (TEL-EH-SKO-PEE-UM) | *The Telescope*

- **Abbreviation:** Tel • **Genitive:** Telescopii • **Area:** 252 square degrees • **Size Ranking:** 57th
- **Best Viewed:** July to August • **Latitude:** 33ºN–90ºS • **Width:** • **Height:**

This faint constellation, dedicated to the astronomer's primary tool, was a creation of Nicolas de Lacaille in 1752. None of its stars are brighter than about magnitude +4, and to the naked eye it appears a rather blank region. But it holds a good selection of faint galaxies for eight-inch (20-cm) and larger telescopes.

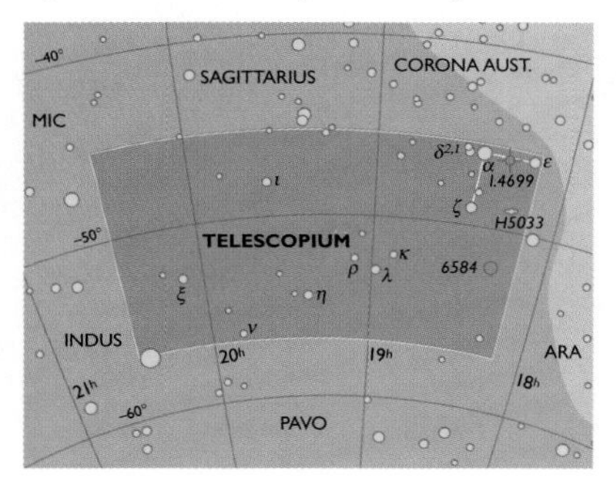

FEATURES OF INTEREST

H 5033: A quadruple star, easily resolved in a small telescope. The primary star is magnitude +7; component B is 17.3 arc seconds distant and magnitude +9.5. Component C is the faintest at magnitude +11, separated by 18.2 arc seconds. The final component, D, is magnitude +9.5 and 27.9 arc seconds away from the primary.
NGC 6584: A conspicuous globular cluster with a magnitude of +8.5, diameter about six arc minutes. It is rather loosely structured, but its brightest stars hover around magnitude +15; at least a twelve-inch (30-cm) telescope is needed to resolve well.

TRIANGULUM (TRI-ANG-GYU-LUM) | *The Triangle*

• **Abbreviation:** Tri • **Genitive:** Trianguli • **Area:** 132 square degrees • **Size Ranking:** 78th

• **Best Viewed:** November to December • **Latitude:** 90ºN–52ºS • **Width:** • **Height:**

Despite its rather small size, this constellation is a conspicuous grouping of stars, and it was well known in ancient times. The Greeks knew the stars as Deltoton because of its resemblance to their capital letter, delta; The region, located away from the plane of the Milky Way, embraces a number of faint galaxies in the region that show up well with telescopes eight inches (20 cm) and larger. There is one Messier object: the Local Group galaxy M33.

M33
This spiral galaxy is the third-largest member of our Local Group of galaxies.

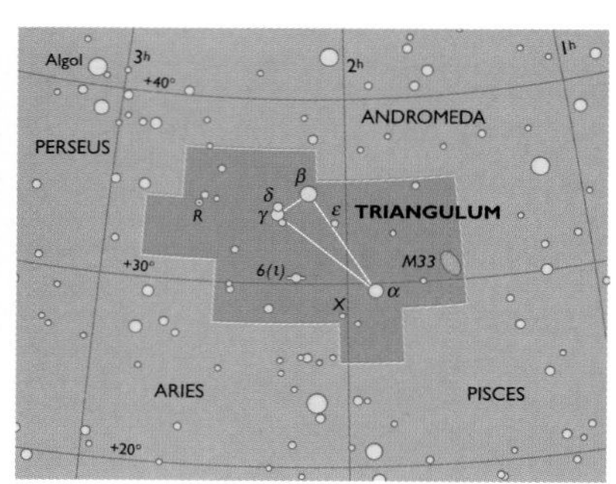

FEATURES OF INTEREST

EPSILON (ε) TRIANGULI: A binary system and a challenge to small telescopes because of the faintness of the secondary. Component A is magnitude +5 star; the B component is magnitude +11 and 3.9 arc seconds distant.

IOTA (ι) TRIANGULI: A pair of yellowish stars, magnitudes +5 and +6.5. The separation is 3.8 arc seconds.

X TRIANGULI: An eclipsing variable with a wide range in brightness in a very short period. The star varies between magnitude +8.9 and +11.5 in less than a day.

MESSIER 33: Under dark skies, this is a good object to view with binoculars, visible as a misty patch of light with poorly defined edges. Small telescopes provide a disappointing view owing to the large size of the galaxy. In twelve-inch (30-cm) and larger telescopes, the spiral structure becomes evident, and a number of faint nebulae can be identified by the careful observer, the easiest of which is NGC 604, an enormous star-forming region. Magnitude: +5.7; distance: 2.2 million light-years.

TRIANGULUM AUSTRALE (TRI-ANG-GYU-LUM OS-TRAY-LEE) | *The Southern Triangle*

• **Abbreviation:** TrA • **Genitive:** Trianguli Australis • **Area:** 110 square degrees • **Size Ranking:** 83rd

• **Best Viewed:** June to July • **Latitude:** 19ºN–90ºS • **Width:** • **Height:**

This far-southern constellation was included in Bayer's *Uranometria* in 1603. The main stars form a roughly equilateral triangle near two of the sun's closest neighbors, Alpha and Beta Centauri.

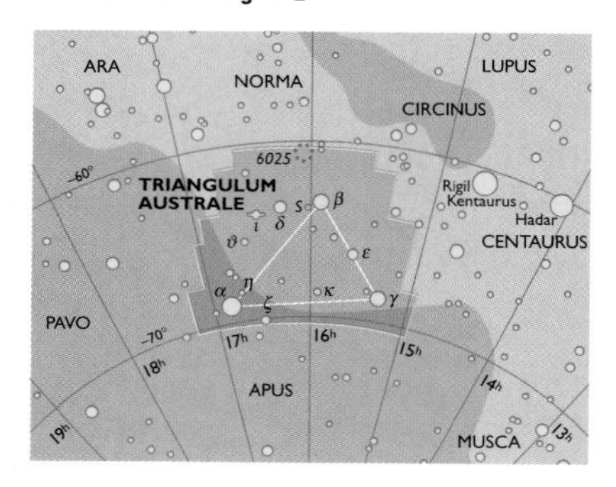

FEATURES OF INTEREST

IOTA (ι) TRIANGULI AUSTRALIS: A wide double star—an optical pair—magnitudes +5.5 and +10. The two are separated by 19.6 arc seconds.

NGC 6025: A bright and easy open cluster, quite elongated in form, featuring about thirty magnitude +7 and fainter stars.

TUCANA (TOO-KAN-AH) | *The Toucan*

• **Abbreviation:** Tuc • **Genitive:** Tucanae • **Area:** 295 square degrees • **Size Ranking:** 48th

• **Best Viewed:** September to November • **Latitude:** 14ºN–90ºS • **Width:** • **Height:**

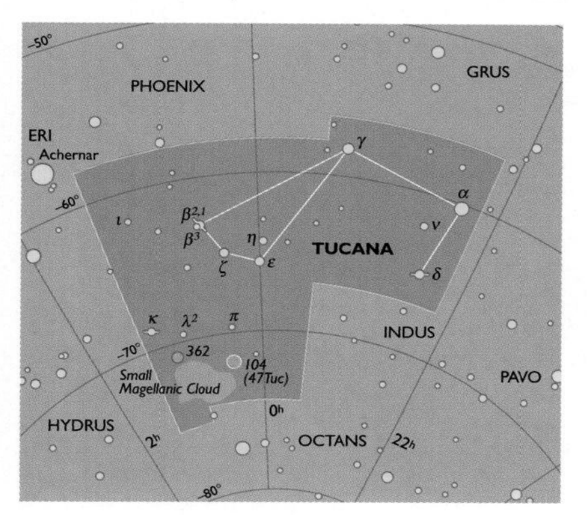

This southern constellation, made up primarily of magnitude +3 to +5 stars, was another of Bayer's creations in 1603. Despite its location far from the plane of the Milky Way, there is a dearth of bright external galaxies. However, Tucana is home to a number of notable objects, including two bright globular clusters and the Small Magellanic Cloud, located near its southern boundary.

FEATURES OF INTEREST

BETA 1 (β1) AND BETA 2 (β2) TUCANAE: A six-star system. Two components are visible with the naked eye—magnitudes +4 and +5; they are separated by twenty-seven arc seconds. A third component, magnitude +5.1, is located 9.3 arc minutes to the southeast. Each of these three stars is a close binary system.

KAPPA (κ) TUCANAE: A double star, with a +5 magnitude component A and a +7 magnitude component B, separated by 5.4 arc seconds.

SMALL MAGELLANIC CLOUD: In both apparent size and brightness, the SMC is definitely inferior to the Large Magellanic Cloud in Dorado, but it is a wondrous sight nonetheless, visible on dark, moonless nights as a bright patch about 3.5 degrees in diameter. There are thirty-one NGC objects in or bordering the galaxy, mostly nebulae and open star clusters. Two of them, NGC 346 and NGC 371, are especially noteworthy as enormous stellar associations, seemingly larger than anything in our own galaxy. The whole galaxy is wonderful to sweep with a telescope, revealing field after field of resolved star dust.

NGC 104: Also known as 47 Tucanae, this enormous globular cluster is second in size and brightness only to Omega Centauri in our galaxy. Rather compressed at its core, the galaxy stretches about twenty-five arc minutes in diameter. Its brightest stars are about magnitude +11, so resolution is possible with a three- or four-inch (8- or 10-cm) telescope. Magnitude: +4.35; distance: about sixteen thousand light-years.

NGC 362: A bright, compressed globular cluster, magnitude +6, with a diameter about ten arc minutes. Resolution is possible with four-to six-inch (10- to 15-cm) telescopes. Distance: about twenty-nine thousand light-years.

The Small Magellanic Cloud

This irregular galaxy, a satellite of the Milky Way, has a visible face at least ten thousand light-years across. It is located about 195,000 light- years from Earth.

URSA MAJOR (ER-SUH MAY-JER) | *The Great Bear*

• **Abbreviation:** UMa • **Genitive:** Ursae Majoris • **Area:** 1,280 square degrees • **Size Ranking:** 3rd

• **Best Viewed:** February to May • **Latitude:** 90ºN–16ºS • **Width:** • **Height:**

This constellation features the best-known asterism in the heavens, the seven stars that make up the Big Dipper. The constellation has been recognized by all ancient cultures as representing the form of a bear. In Greek mythology, Zeus attempted to save the nymph Callisto from the jealous Hera by transforming her into a bear. But Hera was relentless in pursuit, and prevailed upon the goddess of the hunt, Artemis, to slay the bear, whereupon Zeus translated its form to the stars. This enormous constellation contains seven Messier objects within its confines and many variable and binary stars.

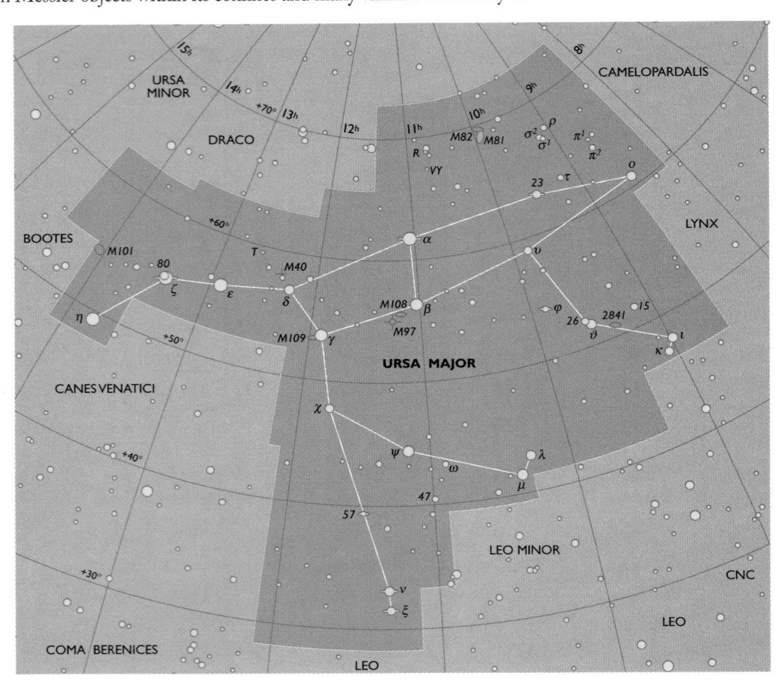

FEATURES OF INTEREST

ZETA (ζ) URSAE MAJORIS (MIZAR): With the nearby star Alcor, Mizar forms a naked-eye double star in the middle of the Big Dipper's handle. Zeta is itself a double, the first to be discovered in the telescopic age, in 1650. Magnitudes: +2.27 and +3.94; separation: 14.4 arc seconds.

XI (ξ) URSAE MAJORIS: Yellowish binary stars that orbit each other every sixty years. Magnitudes +4.3 and +4.8, separated by 1.8 arc seconds.

MESSIER 81 AND MESSIER 82: M81 is a classic Sb-class spiral, appearing as a large, bright-centered oval glow in a small telescope. Magnitude: +6.8. Messier 82 is 0.5 degrees away, an irregular galaxy seen edge on with an odd perpendicular dust lane, visible in six-inch (15-cm) and larger telescopes, bisecting its core. Magnitude +8.4.

MESSIER 97 AND MESSIER 108: This so-called odd couple consists of a large, pale planetary nebula (M97, magnitude: 11.2) and an edge on spiral galaxy (M108, magnitude: 10). M97 is also known as the Owl Nebula because of two dark areas in its disk, visible with a twelve-inch (30-cm) telescope.

M81
Situated twelve million light-years from Earth, M81 is a stunning sight with binoculars or a small telescope.

URSA MINOR (ER-SUH MY-NER) | *The Lesser Bear*

• **Abbreviation:** UMi • **Genitive:** Ursae Minoris • **Area:** 256 square degrees • **Size Ranking:** 56th
• **Best Viewed:** May to June • **Latitude:** 90ºN–0ºS • **Width:** • **Height:**

This is one of the most famous constellations in the heavens, embracing the Little Dipper with Polaris representing the beginning of its handle. There is not much of interest here for the amateur astronomer with binoculars or a small telescope. Ursa Minor includes several faint galaxies, but these typically require telescopes with apertures of twelve inches (30 cm) and up. Only three stars, Polaris, Beta (β), and Gamma (γ) Ursae Minoris are likely to be seen with the naked eye by urban or suburban observers.

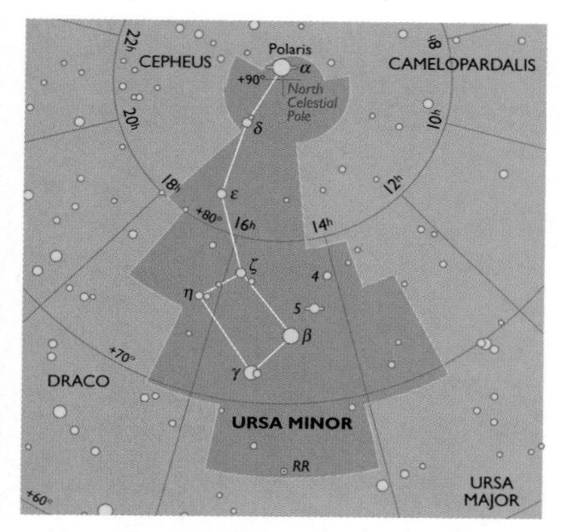

FEATURES OF INTEREST

ALPHA (α) URSAE MINORIS: This is Polaris, the Pole Star. During the night, the entire heavens in the northern hemisphere appear to revolve around it. Polaris is slowly approaching true north, and will be closest to the pole in the year 2102. Users of three-inch (8-cm) and larger telescopes will note a magnitude +9 companion to Polaris about 18.5 arc seconds distant. These two stars are gravitationally bound and are separated by 186 billion miles (299,000,000,000 km). In the small gap visible between them, more than twenty-five solar systems could be placed end to end.

GAMMA (γ) URSAE MINORIS: A double star with a blue-white giant, magnitude 3.0, and an orange giant companion of magnitude 5.

VELA (VEE-LAH) | *The Sail*

• **Abbreviation:** Vel • **Genitive:** Velorum • **Area:** 500 square degrees • **Size Ranking:** 32nd
• **Best Viewed:** February to April • **Latitude:** 32ºN–90ºS • **Width:** • **Height:**

Vela, the Sail, was part of the great ancient constellation Argo Navis that was divided up by Nicolas de Lacaille in the 1750s into four smaller constellations. Rich star fields of the Milky Way traverse this area of the sky from northwest to southeast. This is a good area to hunt for open star clusters, especially with binoculars.

FEATURES OF INTEREST

GAMMA (γ) VELORUM: A bright double star, magnitudes +1.88 and +4.5, separated by forty-one arc seconds. The primary is a Wolf-Rayet star: an extremely hot, highly luminous star that is losing its mass at a fast rate.

NGC 2547: A bright, scattered star cluster with members that are magnitudes +7 and fainter.

NGC 2736: A bright, pencil-thin nebula, the eastern portion of the Vela Supernova Remnant, a star that exploded twelve thousand years ago. You'll need at least an eight- to ten-inch (20- to 25-cm) telescope to see NGC 2736 clearly.

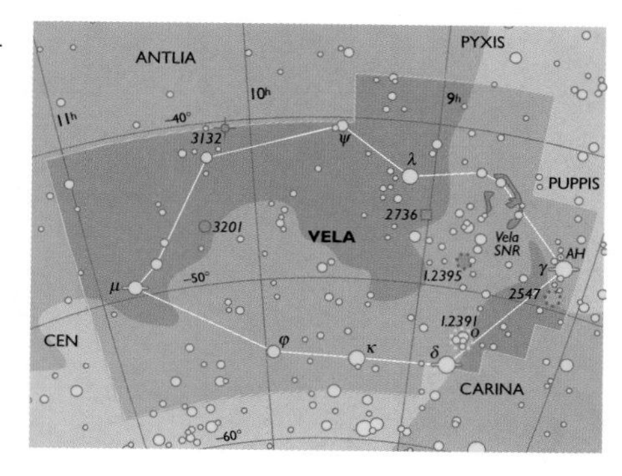

NGC 3132: Called the "Eight-burst" nebula for its figure-eight pattern that appears in long-exposure photos. It is visible with three-inch (8-cm) telescopes. Magnitude: +8.2; diameter: thirty arc seconds.

IC 2391: A bright, coarse cluster, suitable for binoculars.

IC 2395: A small cluster of about fifteen to twenty magnitude +9 and fainter stars.

VIRGO (VER-GO) | *The Virgin*

• **Abbreviation:** Vir • **Genitive:** Virginis • **Area:** 1,294 square degrees • **Size Ranking:** 2nd

• **Best Viewed:** April to June • **Latitude:** 67ºN–75ºS • **Width:** ✋✋✋ • **Height:** ✋✋

This is the largest of the zodiac constellations. Virgo was associated with Astraea, the Greek goddess of Justice and the last of the deities to leave Earth and go to heaven when the sinfulness of man proved intolerable. Located far from the plane of our galaxy, Virgo is notable for containing the brightest, nearest galactic supercluster, the Virgo Cluster. The constellation embraces hundreds of galaxies that can be located with an eight-inch (20-cm) telescope. It is the home of eleven Messier objects, all galaxies.

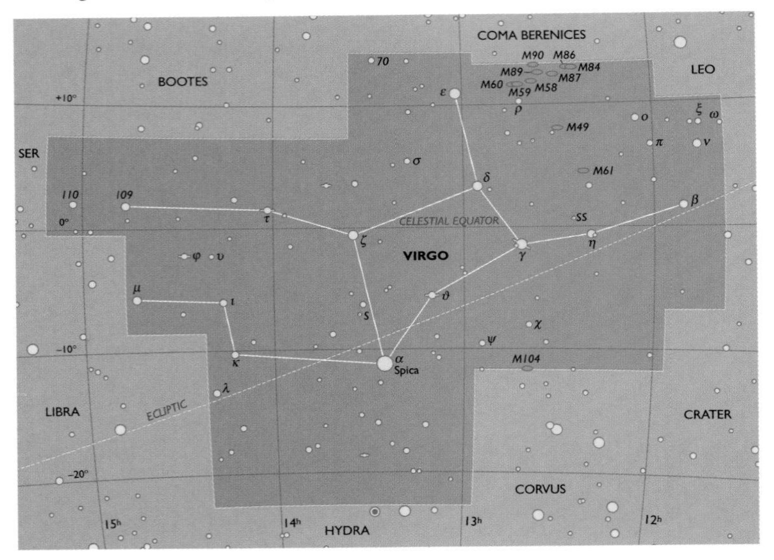

FEATURES OF INTEREST

ALPHA (α) VIRGINIS (SPICA): A blue-white star of magnitude +1, Spica is one of the twenty brightest stars in the sky.

GAMMA (γ) VIRGINIS: A binary star with two yellow-white stars that orbit each other every 170 years. Magnitudes: +3.48 and +3.5; separated by 1.5 arc seconds, a distance that will close until 2005, at which point the two will begin to move apart.

MESSIER 84, 86, 87, 89: These four elliptical galaxies are located within two degrees of each other in the heart of the Virgo Cluster. Messier 87 is particularly noteworthy as a supermassive assemblage of more than one trillion stars. Messier 84 and 86 can be seen together with a low-power eyepiece. The four Messier's range in magnitude from +8.6 to +9.8.

MESSIER 58, 59, 60: These three galaxies are located along an east-west line 1.5 degrees wide. M59 and M60 are bright ellipticals, while M58 is a bright, SB-type barred-spiral. Magnitudes: M58: +8.8; M59: +8.8; M60: +8.8.

MESSIER 49: A large, bright elliptical galaxy similar to M87. Magnitude: +8.4.

MESSIER 90: A bright, classic Sb-type spiral galaxy, with a bright core. Magnitude: +9.5.

MESSIER 104: Also known as the "Sombrero Galaxy" for its large, bright core and tightly-wound, dusty spiral arms that are reminiscent of a hat brim. The dust lane is visible with twelve-inch (30-cm) and larger telescopes. Magnitude: +8.3.

M104
The "brim" of the Sombrero Galaxy is actually a dust lane of dark interstellar matter.

VOLANS (VOH-LANZ) | *The Flying Fish*

• **Abbreviation:** Vol • **Genitive:** Volantis • **Area:** 141 square degrees • **Size Ranking:** 76th
• **Best Viewed:** January to March • **Latitude:** 14ºN–90ºS • **Width:** • **Height:**

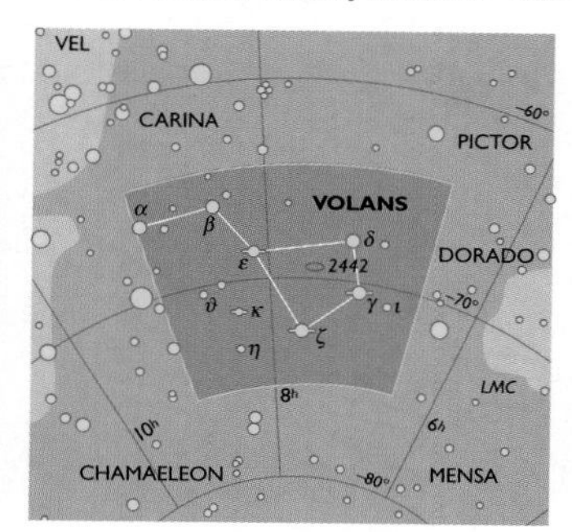

This small constellation was one of Bayer's creations for his *Uranometria* star atlas of 1603. Eight of its stars hover around fourth magnitude and there are a few interesting double stars, along with one galaxy of interest.

FEATURES OF INTEREST

GAMMA (γ) VOLANTIS: A double star with a magnitude +4 component A and a +5.5 component B, separated by 13.6 arc seconds.

EPSILON (ε) VOLANTIS: Another double star. The primary is magnitude +4.5, while the secondary, located 6.1 arc seconds distant, is magnitude +8.

NGC 2442: An eight-inch (20-cm) telescope will show this SBb-type galaxy as an elongated haze with a brighter core. Larger telescopes will reveal something of its spiral pattern. Magnitude: +10.4.

VULPECULA (VUL-PECK-YOU-LAH) | *The Fox*

• **Abbreviation:** Vul • **Genitive:** Vulpeculae • **Area:** 268 square degrees • **Size Ranking:** 55th
• **Best Viewed:** August to September • **Latitude:** 90ºN–61ºS • **Width:** • **Height:**

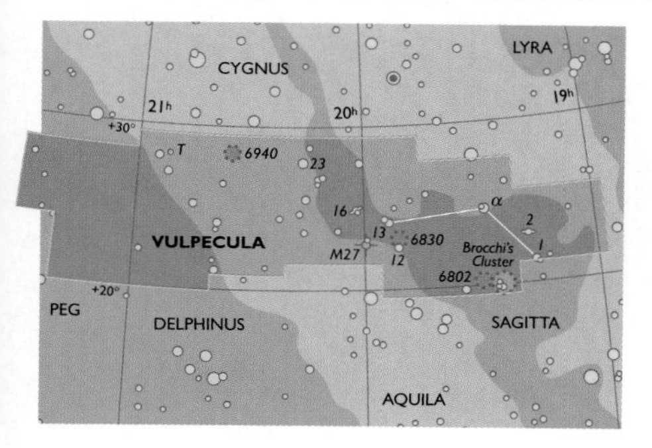

This faint, scattered group of stars is somewhat elusive due to its location in a bright part of the Milky Way, but the Polish astronomer Hevelius saw fit to create a constellation here in the year 1660. Despite being inconspicuous to the naked eye, this is an interesting patch of sky for binoculars or a small telescope. The constellation contains one Messier object, the planetary nebula M27.

FEATURES OF INTEREST

COLLINDER 399: This is also known as Brocchi's Cluster, or more prosaically, the "Coathanger" cluster, a name inspired by its striking resemblance to this Earthly object. Collinder 399 is a bright association of stars, perfect for binoculars. Users of six-inch (15-cm) and larger telescopes will notice the faint open cluster NGC 6802 next to the Coathanger's easternmost star, 7 Vulpeculae. Individual stars can be detected with eight-inch (20-cm) and larger instruments.

MESSIER 27: This is probably the finest specimen of a planetary nebula for small-telescope users, owing to its large size, brightness, and distinctive structure. Dubbed the "Dumbbell" nebula for its oblong shape with wide, rounded extremities, M27 can be seen with a three-inch (8-cm) telescope, although the full extent of the nebular is only revealed with fourteen-inch (36-cm) and larger telescopes. Magnitude: +8.1.

NGC 6830: A small cluster of magnitude +11 and fainter stars that show some compression toward the center.

NGC 6940: This open cluster features a half-dozen magnitude +9 stars overlaying a grouping of fine stardust. Telescopes six inches (15 cm) and larger will reveal from sixty to one hundred stars.

Astronomer's Resource Guide

This guide includes a list of some of the extremes of astronomy (opposite); *a simple Sun viewer that you can make from a shoe box* (page 178); *major milestones in astronomy* (page 179); *a list of clubs and associations, planetariums, and Web sites* (pages 180 to 181); *and a glossary* (pages 182 to 183).

Astronomy's Outer Limits

LARGEST TELESCOPES

- Refracting: Yerkes Observatory, 40 inches (1 m), completed in 1897, Williams Bay, WI
- Reflecting: W.M. Keck Telescopes, 390 inches (10 m), Mauna Kea, HI
- Radio: Arecibo, 1,000 feet (300 m), Puerto Rico

LARGEST/ SMALLEST PLANETS (DIA.)

- Jupiter: 88,865 miles (143,014 km)
- Pluto: 1,429 miles (2,300 km)

HOTTEST / COLDEST PLANETS

- Venus: 899°F (482°C)
- Pluto: -387°F (-233°C)

NEAREST STARS

• Proxima Centauri	4.2 light-years
• Centauri A/Centauri B	4.3 light-years
• Barnard's Star	5.9 light-years
• Wolf 359	7.6 light-years
• Sirius A	8.6 light-years
• Sirius B	8.7 light-years

BRIGHTEST STARS: THEIR MAGNITUDES

• Sirius	-1.46
• Canopus	-0.72
• Alpha Centauri	-0.29
• Arcturus	-0.06
• Vega	+0.04
• Capella	+0.08
• Rigel	0.11
• Procyon	0.37
• Achernar	0.46

The Very Large Telescope in Cerro Paranal, Chile, scheduled for completion in the year 2000, will boast four twenty-six-foot (8-m) mirrors, making it the world's largest reflecting telescope. Here, one of the telescopes is shown on its partially completed five-hundred-ton altazimuth mount.

This image of Ganymede—the solar system's largest satellite—was taken from a distance of three thousand miles (4,800 km) by Voyager 1 *on March 4, 1979. The surface shows evidence of intense meteoric bombardment.*

LARGEST SATELLITES IN SOLAR SYSTEM (DIAMETER)

• Ganymede/Jupiter	3,270 miles (5,268 km)
• Titan/Saturn	3,201 miles (5,150 km)
• Callisto/Jupiter	2,983 miles (4,806 km)
• Io/Jupiter	2,256 miles (3,642 km)
• Moon/Earth	2,160 miles (3,476 km)
• Europa/Jupiter	1,950 miles (3,130 km)
• Triton/Neptune	1,678 miles (2,706 km)

Build Yourself a Sun Viewer

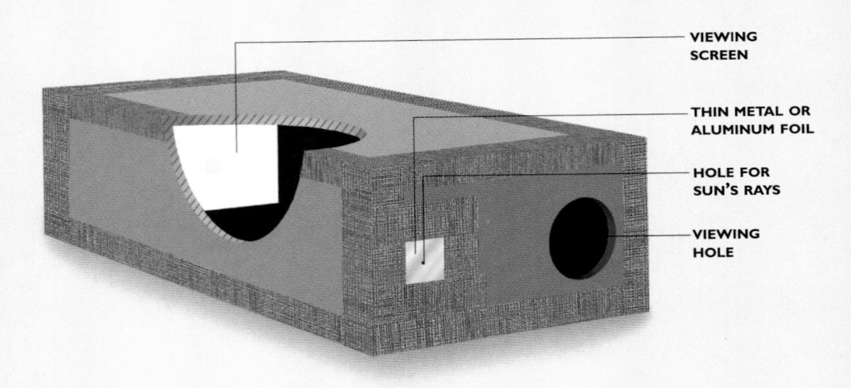

Looking at the Sun is dangerous—especially with a telescope that doesn't have the proper protection *(page 75)*. You can outfit your instrument with special solar filters fitted to the front of the telescope or use a projection screen. A simpler way to observe the Sun is with the homemade viewer shown above.

Here's what you'll need to build it: a cardboard box; some duct tape; a small piece of metal or aluminum foil; black spray paint; a piece of white cardboard; and some glue. The dimensions of the box aren't important. Something the size of a shoebox—roughly fourteen inches long, seven inches wide, and four inches high—will work well. Boxes that are much longer will produce a bigger image of the Sun but it will be more diffuse, while shorter boxes will create a smaller, more intense image.

Spray paint the inside of the box black, except for the end that will be the viewing screen. Glue a piece of white cardboard to cover that part of the box. Cut a one-and-one-half-inch-diameter hole near one side of the opposite end of the box. Then make a small hole for the incoming Sun's rays near the other side of the same end, as shown at left. To do this, cut out a one-inch square from the box and tape a piece of thin sheet metal or several layers of aluminum foil to cover the hole. Then pierce the metal with a large pin. This will create a crisper hole than you would get by punching directly through the cardboard.

Finally, seal the box with duct tape to keep any extraneous light from entering.

To use the viewer, stand with your back to the Sun on a bright day. Position the box so the pinhole faces the Sun. Look through the viewing hole and adjust the position of the box until an image of the Sun appears on the viewing screen at the far end of the box.

MAJOR MILESTONES IN ASTRONOMY

- **c.4200 B.C.** Egyptians develop a 365-day calendar based on the cycles of the Sun and the Moon.
- **763 B.C.** Babylonians first record a solar eclipse.
- **530 B.C.** Greek astronomer Pythagoras suggests that the Earth is spherical.
- **400 B.C.** Babylonians identify signs of the zodiac.
- **350 B.C.** Greek astronomer Heraclides suggests that the Earth rotates on its own axis.
- **240 B.C.** Eratosthenes of Cyrene calculates Earth's circumference within a few hundred miles of its actual measurement: twenty-five thousand miles (40,000 km).
- **Circa 200 B.C.** Greek astronomer Aristarchus proposes that the Earth revolves around the Sun. He also suggests that day and night are caused by the Earth's rotation.
- **A.D. 140** Egyptian astronomer Ptolemy refines the theory that the Earth is the center of the universe,
- **1100** Chinese astronomers explain the cause of eclipses.
- **1440** German scholar Nicholas of Cusa hypothesizes that space is infinite, and that the stars are suns with their own inhabited planets.
- **1514** Polish astronomer Nicolas Copernicus asserts that the Earth and other planets revolve around the Sun.
- **1608** Hans Lippershey, Dutch optician, builds the first telescope.
- **1609** Johannes Kepler proposes the orbital laws of planetary motion.
- **1611** Galileo Galilei publishes *Siderius Nuncius*, confirming the theories of Copernicus. For this, Galileo is tried and condemned by the Inquisition in Rome.
- **1675** Olaus Roemer measures the speed of light.
- **1687** Isaac Newton publishes his law of universal gravitation.
- **1682** British astronomer Edmond Halley observes the predicted return of the comet that will bear his name.
- **1781** William Herschel discovers Uranus.
- **1784** French astronomer Charles Messier publishes a catalog of star clusters and nebulae.
- **1814** Joseph von Fraunhofer begins studying absorption lines in the Sun's spectrum.
- **1838** Friedrich Bessel is the first astronomer to measure the distance to a star—61 Cygni.
- **1888** J.L. Dreyer publishes the New General Catalog (NGC) of stars and galaxies.
- **1905** Albert Einstein publishes his special theory of relativity.
- **1908** Ejnar Hertzsprung revolutionizes the study of stellar evolution by relating the size, brightness, and temperature of stars to their age.
- **1916** Einstein publishes his general theory of relativity.
- **1918** American astronomer Harlow Shapley determines the dimensions of the Milky Way.
- **1927** Georges-Henri LeMaitre develops the Big Bang theory of the origin of the universe.
- **1929** American Edwin Hubble proves that the universe is expanding.
- **1930** American astronomer Clyde Tombaugh discovers Pluto.
- **1931** American radio engineer Karl Jansky detects radio waves originating from outer space.
- **1948** Two-hundred-inch Hale telescope begins operation in California.
- **1959** *Luna 3* takes photos of the never-before-seen far side of the moon.
- **1963** First quasar discovered.
- **1964** American physicists Arno Penzias and Robert Wilson detect the microwave background radiation from the Big Bang.
- **1965** Cygnus X-1, an intense X-ray source, is discovered. It is now believed to be a black hole.
- **1996** Astronomers discover two planets that are part of another solar system.
- **1997** *Mars Pathfinder* expedition lands a robot explorer on Mars.
- **1998** Supermassive black hole confirmed at the center of the Milky Way galaxy.

Important Addresses

ASSOCIATIONS

There are hundreds of amateur astronomy clubs in North America. To find one near you, visit Sky & Telescope's Web page *(opposite)* or contact the associations marked* below.

Astronomical League*
2112 Kingfisher Lane E.
Rolling Meadows, IL 60008
(847) 398-0562
jastevens@compuserve.com
http://astroleague.org

Astronomical Society of the Pacific*
390 Ashton Ave.
San Francisco, CA 94112
(415) 337-1100
asp@stars.sfsu.edu
http://www.aspsky.org

International Dark Sky Association
3225 N. First Ave.
Tucson, AZ 85719
ida@darksky.org
http://www.darksky.org

The Planetary Society
65 North Catalina Ave.
Pasadena, CA 91106-2301
(626) 793-5100
tps@mars.planetary.org
http://www.planetary.org

The Royal Astronomical Society of Canada
136 Dupont St.
Toronto, ON M5R 1V2
Canada
(416) 924-7923
rasc@rasc.ca
http://www.rasc.ca

PLANETARIUMS & OBSERVATORIES

There are scores of planetariums and observatories throughout North America. We have chosen some of the best-known examples. In the case of planetariums, that means ones that can seat at least 225 people. For the name of the planetarium or observatory nearest you, consult *Sky and Telescope's* Web page *(opposite)*.

Adler Planetarium
1300 South Lake Shore Dr.,
Chicago, IL 60605
(312) 322-0590
http://www.adlerplanetarium.org

Arecibo Observatory
Cornell University, P.O. Box 995, Arecibo, PR 00613
(787) 639-4635

Burke Baker Planetarium
One Herman Circle Dr.,
Houston, TX 77030-1799
(713) 639-4635
http://www.hmns.mus.tx.us/

David Dunlop Observatory
Box 360
Richmond Hill, ON L4C 4Y6
(905) 884-2112
http://astro.utoronto.ca/ddohome.html

Albert Einstein Planetarium
(National Air and Space Museum), Independence Ave. between 4th and 7th, Washington, DC
(202) 357-1686
http://www.nasm.si.edu

Fels Planetarium
222 N. 20th St.,
Philadelphia, PA 19103
(215) 448-1208

Fernbank Science Center
156 Heaton Park Dr. NE,
Atlanta, GA 30307
(404) 378-4311
http://www.fernbank.edu/

W.A. Gayle Planetarium
1010 Forest Ave.
Montgomery, AL 36106
(334) 241-4799
http://www.tsum.edu

Griffith Park Observatory and Planetarium
2800 E. Observatory Rd.
Los Angeles, CA 90027
(323) 664-1191
http://www.GriffithObs.org

Hansen Planetarium
15 S. State St., Salt Lake City, UT 84111-1590
(801) 538-2098

Hayden Planetarium
(scheduled to reopen in the year 2000)
81st St. and Central Park West, NYC, NY 10024
Tel: (212) 769-5900
http://www.amnh.org

Charles Hayden Planetarium
Science Park
Boston, MA 02114
(617) 723-2500
http://www.mos.org

Kelly Planetarium
301 N. Tryon St.
Charlotte, NC 28202
(704) 372-6261

Kitt Peak National Observatory
P.O. Box 26732
Tucson, AZ 85726
(520) 318-8200
http://www.noao.edu

Lick Observatory
P.O. Box 85
Mount Hamilton,
CA 95140
(408) 274-5061
http://www.ucolick.org

Lowell Observatory
1400 W. Mars Hill Rd..
Flagstaff, AZ 86001-4499
(520) 774-2096
http://www.lowell.edu

H. R. MacMillan Planetarium
(Pacific Space Centre)
1100 Chestnut St.,
Vancouver, B.C. V6J 3J9
Tel: (604) 738-7827

ASTROLab du Mont Megantic
189 route du Parc
Notre-Dame-des-Bois, Qc
(819) 888-2941

Miami Space Transit Planetarium
3280 S. Miami Ave.
Miami, FL 33129
(305) 854-4244
http://www.starhustler.com/

Mauna Kea Observatory
Onizuka Visitors Information Station
177 Makaala St.
Hilo HI 96720
(808) 961-2180

Morehead Planetarium
(University of North Carolina at Chapel Hill)
250 E. Franklin, NC
(919) 962-1236
http://www.unc.edu/depts/mhplanet/

Morrison Planetarium
California Academy of Sciences,
Golden Gate Park,
San Francisco, CA 94118
(415) 750-7127
http://www.calacademy.org/planetarium/.

Mount Wilson Observatory Assoc.
P.O. Box 70076
Pasadena, CA 91117-7076
http:www.mwoa.org

Palomar Observatory
Palomar Mountain,
CA 92060
http://astro.caltech.edu/observatories/palomar/index.html

Planetarium de Montreal
1000 St. Jacques Ouest,
Montreal, Qc. H3C 1G7
(514) 872-4530
http://www.planetarium.montreal.qc.ca

Willard W. Smith Planetarium
Pacific Science Center, 200
Second Ave. N, Seattle, WA
(206) 443-3648

Strasenburgh Planetarium
Rochester Museum and Science Center
657 East Ave.
Rochester, NY 14603
(716) 271-1880
http://www.rmsc.org

Yerkes Observatory
373 W. Geneva St.,
P.O. Box 258
Williams Bay, WI 53191
(414) 245-5555
http://astro.uchicago.edu/yerkes

WEB SITES

Most amateur and professional astronomy clubs and societies have Web sites. Search for terms such as "astronomy" and "star party" in common search engines such as Alta Vista or Metacrawler. Here are a few good sites to begin with:

The Astronomy Cafe
http://www2.ari.net/home/odenwald/cafe.html

Astronomy Magazine
http://www.kalmbach.com/astro/astronomy.html

Astronomical Society of the Pacific
http://www.aspsky.org/

Cosmos in a Computer
http://www.ncsa.uiuc.edu/Cyberia/Cosmos?CosmosCompHome.html

Deepsky Observer's Companion
http://home.global.co.za~auke/index.htm

Stephen Hawking's Universe
http://www.pbs.org/wnet/hawking/html/home.html

Hubble Heritage Project
http://heritage.stsci.edu

National Aeronautics and Space Administration
http://www.NASA.gov
http://www.jpl.nasa.gov

The **SETI League**
http://www.setileague.org

Sky & Telescope
http://www.skypub.com

GLOSSARY OF ASTRONOMICAL TERMS

Absolute magnitude: The measure of a star's brightness at a standard distance of ten parsecs (32.6 light-years).

Absorption lines: Dark spectral lines caused when specific wavelengths of light are absorbed by an intervening cooler gas.

Accretion disk: A disk of material that gravitates around a star or black hole before gradually falling into it.

Active galaxy: A highly energetic galaxy powered by a supermassive black hole at its core.

Altitude: An object's distance above the horizon, measured in degrees.

Arc minute: One sixtieth of one degree. Each arc minute, in turn, contains sixty arc seconds.

Asterism: A notable pattern of stars, such as the Big Dipper, forming part of a constellation.

Asteroids: Small rocky objects that orbit the Sun, mainly between Mars and Jupiter.

Astronomical unit: Unit of measurement, abbreviated AU, that is the distance between Earth and the Sun: ninety-three million miles (150,000,000 km).

Azimuth: A celestial object's distance from true north measured eastward in degrees along the horizon.

Big Bang: The theory that cosmic expansion began about fifteen billion years ago from a single point.

Binary star: Two stars that are orbiting one another.

Black hole: An object so dense that not even light can escape its gravitational pull.

CCD: Charge-coupled device. A silicon wafer that converts light into an electrical current, which in turn produces an image.

Celestial equator: The projection of the Earth's equator onto the celestial sphere.

Celestial poles: The projection of Earth's north and south polar axes onto the celestial sphere.

Celestial sphere: An imaginary hollow sphere surrounding the Earth on which the stars appear to be fixed.

Cepheid variable: A class of variable star, similar to Delta Cephei, which pulsates in brightness. The brighter the star, the more slowly it pulsates.

Chandrasekhar limit: The mass limit at which a stellar core cannot remain a white dwarf. A stellar core greater than 1.4 solar masses must become a neutron star or a black hole.

Chromosphere: The layer of a star's atmosphere that is just above its photosphere.

Circumpolar stars: The stars around the celestial poles that never set at the observer's particular latitude.

Closed universe: A universe with sufficient mass to gravitationally reverse its expansion and eventually collapse.

Collimation: The alignment of lenses or mirrors in a telescope.

Conjunction: The alignment of celestial objects along the observer's line of sight.

Constellation: One of eighty-eight recognized star figures, now defined as areas of sky that bound these figures.

Corona: The outermost region of a star's atmosphere.

Dark matter: Invisible, non-radiant matter that makes up 90 percent of the universe. It can't be seen, but its presence can be inferred from its gravitational effect on galaxies and galactic clusters.

Declination: The distance of a celestial object above or below the celestial equator, measured in degrees.

Double star: Two stars that orbit one another or, in the case of an optical double, two unrelated stars in the same line of sight.

Dwarf star: A small main sequence star like the Sun, or in the case of a white dwarf, the remnant core of a Sunlike star. A brown dwarf is a star with insuficient mass to initiate a fusion reaction.

Eclipse: The passage of a celestial body into the shadow of another. Solar eclipses occur when the Moon's shadow is cast on Earth; lunar eclipses occur when the Moon passes through Earth's shadow.

Ecliptic: The plane of planetary orbits projected onto celestial sphere. Also the Sun's apparent annual path across the sky.

Electromagnetic spectrum: The full range of electromagnetic radiation, from long radio waves to short wavelength gamma rays.

Elongation: A planet's distance east or west of the Sun, measured in degrees.

Emission lines: Bright spectral lines caused when specific wavelengths are emitted by a hot glowing gas.

Emission nebula: A glowing cloud of interstellar gas.

Event horizon: The boundary of no return around a black hole, beyond which no event can be observed from outside.

Flat universe: A universe with enough mass to slow down its expansion without ever stopping.

Galactic disk: Flat disk of stars, dust, and gas in spiral galaxies.

Galactic halo: A vast spherical region of old stars and globular star clusters around a galaxy.

Galaxy: A vast stellar island held together by gravity; may be irregular, elliptical, barred, or regular spiral.

Globular cluster: An ancient, spherical cluster of up to a million stars in orbit around galaxies.

Gravitational lens: The bending of radiation from a distant source by the gravitational field of an intervening mass such as a galaxy.

Index Catalogue (IC): A supplement to the NGC. The IC contains 5,386 star clusters, nebulae, and galaxies.

Inflationary-era theory: A theorized brief initial phase of the Big Bang in which the universe was driven to expand faster than the speed of light by a repulsive force contained in the primordial vacuum of empty space.

Interferometry: The use of more than one telescope to increase resolution by combining light or radio waves coming from the same object.

Kelvin: A temperature scale based on absolute zero, -460°F / -273.15°C, at which the motion of molecules stops. (0°C = 273.15 Kelvins.)

Kuiper Belt: A disk of icy objects beyond Pluto's orbit that merges with the Oort Cloud; repository for short-period comets.

Light-year: The distance that light travels in one year: six trillion miles (9,470,000,000,000 km).

Local group: A gravitationally bound cluster of about thirty local galaxies that includes the Milky Way.

Magnitude: The brightness of a celestial object. Each decreasing magnitude is 2.51 times brighter than the last.

Main sequence: The main part of a star's life cycle in which it under-goes hydrogen fusion.

Meridian: An imaginary line that bisects the sky and runs from north to south through the zenith.

Messier Catalog (M): A catalog of 110 bright objects, including star clusters, nebulae, and galaxies.

Mira star: One of a class of variable stars whose luminosity varies over a long period of time. Its prototype is Mira in constellation Cetus.

Nebula: An interstellar cloud of dust and gas.

Neutron star: A massive degenerate stellar core twenty miles (32 km) across that is so dense its protons and electrons fuse together forming neutrons.

NGC: New General Catalog. A catalog of 7,840 star clusters, nebulae, and galaxies.

Nova: A star that flares in brightness when it draws gas from a companion star.

Occultation: The hiding of an object such as a star by a nearby object such as the Moon.

Oort Cloud: A vast spherical halo of icy objects around the solar system, the repository of long-period comets.

Open cluster: A loose association of recently formed stars.

Open universe: A universe with insufficient mass to gravitationally slow its expansion.

Opposition: The point in a planet's orbit when the planet is directly opposite the Sun, as seen from Earth.

Parallax: The apparent shift of a foreground object against the distant background when seen from different angles.

Parsec: Parallax second. The distance at which the Earth appears one arc-second away from the Sun (3.26 light-years).

Photosphere: The visible surface of a star.

Planetary nebula: The outer shell of gas ejected by a red giant prior to becoming a white dwarf.

Precession: A 25,800-year wobbling cycle of the Earth's rotation axis that causes a gradual shift of the celestial poles and celestial coordinates.

Prominence: A magnetically contained loop of solar material ejected above the Sun's surface.

Proper motion: The apparent motion of a star across the celestial sphere, as seen from Earth.

Pulsar: A rotating neu-tron star with a magnetically focused radio beam that creates radio pulses as it sweeps past Earth.

Quasar: Quasi-stellar object. The extremely energetic early stage in the formation of an active galaxy, characterized by a prodigious output of radiation.

Radio galaxy: An active elliptical galaxy that emits much of its radiation at radio wavelengths.

Red shift: A shift of spectral lines toward the red end of the spectrum, caused by the stretching of wavelengths due to a celestial object's motion away from Earth.

Retrograde motion: The apparent reversal of the normal west-to-east motion of outer planets as the Earth passes them.

Right ascension: The celestial equivalent to lines of longitude, which run though the celestial poles and are perpendicular to the celestial equator: They are scaled in hours, increasing eastward from 0h and going full-circle to 24h.

Seyfert galaxy: An active spiral galaxy powered by a black hole; probably a late stage of quasar evolution.

Sidereal time: A time scale based on the true period of Earth's rotation: the time from star rise to star rise (23h, 56m, 4s).

Singularity: An infinitely dense point at the center of a black hole.

Solar wind: An outflow of charged sub-atomic particles streaming from the Sun.

Spacetime: The unified four-dimensional frame-work of our universe (three dimensions of space and one of time) curving under the influence of mass.

Spectral types: The classification of stars into types O, B, A, F, G, K, and M based on temperature and color (type O: hot and blue; type M: cool and red).

Sunspot: A relatively cool dark area where magnetic fields pierce the solar surface.

Supercluster: An immense association of thousands of galaxies that are bound together by gravity.

Supernova: Massive explosion in which a star blows itself apart.

Variable star: A star with luminosity that varies intrinsically or due to external influences.

Zodiac: A band of sky extending nine degrees on each side of the ecliptic. The zodiac contains the twelve zodiac constellations against which the Sun, Moon, and planets—with the exception of Pluto—move.

Index

Text references are in plain type; illustrations and constellation maps in *italic*; photographs in **bold**; and full descriptions of constellations in *italic* with asterisk (*). Entries beginning with numbers appear at the beginning of the index.

St. Remy Media

President: Pierre Léveillé
Vice-President, Finance: Natalie Watanabe
Managing Editor: Carolyn Jackson
Managing Art Director: Diane Denoncourt
Production Manager: Michelle Turbide
Director, Business Development: Christopher Jackson
Senior Editor: Pierre Home-Douglas
Art Director: Solange Laberge
Contributing Art Director: Philippe Arnoldi
Writers: Louie Bernstein, Mark Bratton, Neale McDevitt, Kenneth J. Ragan
Night Sky Cartographer: Wil Tirion
Illustrators: François Escalmel, Patrick Jougla
Photo Researcher: Linda Castle
Researcher: Adam Van Sertima
Indexer: Linda Cardella Cournoyer
Senior Editor, Production: Brian Parsons
Systems Director: Edward Renaud
Technical Support: Jean Sirois, Roberto Schulz
Scanner Operators: Martin Francoeur, Sara Grynspan

The following persons also assisted in the preparation of this book:
Éric Archambault, Robert Chartier, Hélène Dion, Lorraine Doré, Alan Dyer, Laird Greenshields and Maryo Proulx.

Acknowledgments

The editors wish to thank the following:
Dr. David Anderson, University of California, Berkeley; Dave Beckett, Laser Interferometer Gravity Wave Observatory; Dr. Ermanno F. Borra, Université Laval; George Constable, for his editorial contributions; Dave Finley, National Radio Astronomy Observatoy; Jason Fournier, Celestron International; Dr. E. Douglas Hallman, Sudbury Neutrino Observatory; Eric Kopit, Orion Telescopes & Binoculars; Ellis D. Miner, JPL Public Affairs; Mary Beth Murrill, JPL Public Affairs; Mark Robinson, Northwestern University; Gary Sanders, Laser Interferometer Gravity Wave Observatory; Toni Stevens, Orion Telescopes & Binoculars; Wallace Tucker, Chandra X-Ray Observatory Public Affairs; Dan Wertheimer, University of California, Berkeley; and Kisha Wright, Goddard Space Flight Center Public Affairs.

Picture Credits

NOAO/TSADO/Tom Stack & Associates-6-7, 121 (upper), 134, 163 (lower); Tony Ward, Tetbury/Science Source/Photo Researchers-10 (upper), 109 (upper); NASA/Science Source/Photo Researchers-12 (lower), 17, 36, 37 (lower), 40, 47 (upper), 177 (upper); Frank Zullo/Science Source/Photo Researchers-20, 64-65; Celestial Image Co./Science Source/Photo Researchers-22, 23, 169; Royal Observatory, Edinburgh/AATB/Science Source/Photo Researchers-26, 31; Science Source/Photo Researchers-34 (upper), 48 (lower); European Space Agency/Science Source/Photo Researchers-34 (lower); National Space Science Data Center-35, 39 (both), 43, 44-45, 50-51, 56; JPL/TSADO/Tom Stack & Associates-37 (upper); ESA/TSADO/Tom Stack & Associates-38; U.S. Geological Survey/Science Source/Photo Researchers-41 (upper); USGS/TSADO/Tom Stack & Associates-41 (lower); NASA/JPL/Tom Stack & Associates-46; Pekka Parviainen/Science Source/Photo Researchers-48 (upper); Detlev Van Ravenswaay/Science Source/Photo Researchers-49; Bob Burch/Bruce Coleman Inc.-52-53; David Nanuk/Science Source/Photo Researchers-54, 82; Chip Simons-55 (upper); Richard J. Wainscoat-55 (lower); NASA/Tom Stack & Associates-57(both); Photo Courtesy of TRW Inc.-59 (upper); NASA, Compton Gamma Ray Science Support Center-59 (lower); Courtesy of LIGO Project/California Institute of Technology-60; Courtesy of The Sudbury Neutrino Observatory-61 (lower); David Parker/Science Source/Photo Researchers-62, 177 (lower); NASA/JPL/TSADO/Tom Stack & Associates-63; Robert Chartier-66; Courtesy of Meade Instruments Corporation-68, 69 (upper), 71 (left), 78; Maryo Proulx-69; Courtesy of Orion Telescopes & Binoculars-70, 71 (right), 73 (all); Jerry Lodrigues/Science Source/PhotoResearchers-77; Jason Ware-79; Buzzelli/O'Brine/Tom Stack & Associates-80-81; Jerry Schad/Science Source/Photo Researchers-87; Rex A. Butcher/BUTCH/Bruce Coleman Inc.-90-91; Mary Evans Picture Library-92, 176; UC Regents/Lick Observatory-98-99; USNO/TSADO/Tom Stack & Associates-109 (lower); Bill & Sally Fletcher/Tom Stack & Associates-111, 154, 170; John Sanford/Science Source/Photo Researchers-113, 118, 138; Jack Newton-97, 117, 123 (upper), 124, 126, 128, 130, 131, 133, 135, 141, 142, 145, 149, 151, 153, 155, 156, 165, 172, 174; Noel T. Munford/Science Source/Photo Researchers-120; Ronald Roger/Science Source/Photo Researchers-121 (lower); Kim Gordon/Science Source/Photo Researchers-123 (lower); Mike O'Brine/Tom Stack & Associates-132; NOAO/Tom Stack & Associates-163 (upper); Stocktrek-167; Dr. Luke Dodd/Science Source/Photo Researchers-168, 171.